The WAY
KITCHENS
WRK

WITHDRAWN

The WAY KITCHENS WORK

The Science Behind the Microwave, Teflon Pan, Garbage Disposal, and More

ED SOBEY

CHICAGO
REVIEW
PRESS

Library of Congress Cataloging-in-Publication Data
Sobey, Edwin J. C., 1948–
    The way kitchens work : the science behind the microwave, teflon pan, garbage
disposal, and more / Ed Sobey.
        p. cm.
    Includes bibliographical references (p.   ).
    ISBN 978-1-56976-281-3
    1. Cookery—Equipment and supplies 2. Kitchens—Equipment and supplies.
I. Title.

    TX656.S656 2010
    643'.3—dc22

                                                              2010007250

Cover design, interior design, and interior illustrations: Scott Rattray
Cover photos: iStock.com

Published by Chicago Review Press, Incorporated
814 North Franklin Street
Chicago, Illinois 60610
ISBN 978-1-56976-281-3
Printed in the United States of America
5 4 3 2 1

To Aunt Jean—remembering wonderful meals

CONTENTS

ACKNOWLEDGMENTS

Like several of my other books, *The Way Kitchens Work* benefits from photographs taken by Rich Sidwa. I have learned to ask for his help just after we finish a long run through the Watershed Nature Preserve, when he is feeling good but has weakened resistance.

Dave Wilson of Appliance Recycling Outlet in Snohomish, Washington, allowed me to take photos of appliances being repaired or recycled. He introduced me to Ron Gale, a wonderful guy from Forks, Washington, who took me on a tour of the recycling center and helped me take photographs of several kitchen devices. Ron grew up in the hardware and appliance business in Forks and was a fountain of knowledge for a science writer who had a lot to learn about major appliances.

Frank Slagle contributed his ancient toaster for me to take apart and photograph. They don't make them like this—either Frank or his toaster—anymore.

Metal Ware Corporation, manufacturers of Nesco products, sent the photo of their beautiful coffee roaster. The Fire House in Redmond, Washington, provided me with two fire extinguishers to take apart. Thanks, Christian.

Woody, my son and coauthor on *The Way Toys Work*, provided information about and photographs of homemade coffee roasters. Sue Maybee loaned her electric knife for a photo shoot.

Thank you all.

INTRODUCTION

During dinner at a friend's house, I wait as our host leaves the kitchen to fetch some wine. The hostess is escorting Barbara, my wife, through their recent remodeling project, so I am left on my own.

Then I spot it! I ease the digital camera from my pants pocket and press the power button. Slowly I stalk my unsuspecting target: a salad spinner. Very cool. We don't have one. I check the lighting and snap away. This is one of the many photos I need for this book.

Writing a book on kitchens dictates that I carry a camera with me nearly everywhere I go, and it takes me to Goodwill outlets to find used appliances I can disassemble. With a few screwdrivers, a pair of pliers, and my trusty digital camera, I take apart my thrift-store finds and make all manner of discoveries. Writing this book has brought me much better understanding and great enjoyment. I hope it brings you the same.

What's Cooking?

The kitchen is a magnet for people and a showcase for technology. Everyone gathers there to eat, talk, or just hang out. All around them, kitchen shelves and countertops are home to cutting-edge technology—the latest ways to slice, dice, mix, bake, and cook. And the ideas keep coming; many hundreds of new innovations are patented each year.

Today's wonders will be surpassed tomorrow, just like earlier technology gave way to what we have today. The first kitchen utensils probably consisted of sharp rocks for cutting flesh, heavy rocks for breaking bones to get to the marrow inside, and leaves folded into bags to carry food. Searching kitchens today I haven't yet found a rock tool or leaf. The technology you see in a modern kitchen is the result of 10,000 years of design evolution.

As so often happens, a single innovation fundamentally changes everything. Fire for cooking was one such innovation. No one holds a patent for fire, and most likely thousands of creative people in many locations discovered it independently. People everywhere adopted it, and as they did they generated new challenges for the kitchen. How do you take advantage of the heat without destroying the food or its container, or burning down your home? How do you control the flames, smoke, and ash?

Some experts believe that cooking was first developed about 2 million years ago, while others claim it was invented much more recently. At first, cooking was done over wood fires, and possibly in ovens made of packed dirt. Prehistoric chefs had to go outside of their homes in order to cook; the kitchen as a separate room in an inhabited building was invented long afterward, in the Middle Ages.

About 9,000 years ago, humans developed earthenware pottery for use in baking and the storage of food and beverages. Some 5,300 years ago, they invented bronze, an alloy of copper and tin that could hold a sharpened edge, allowing the introduction of cutting implements. The discovery of iron 2,000 years later vastly improved the quality of knives and other tools.

Ancient Greeks had open-air kitchens in the center of their homes. Most Roman cooks met at communal kitchens to prepare meals, while the wealthy among them had their own kitchens as separate rooms in their mansions. In medieval times, kitchens were located in the center of a home's living area. Families would gather there to enjoy the warmth of the smoky fire as well as each other's company. A hole in the center of the roof let the smoke out. By the 12th century, chimneys were constructed to siphon the smoke up and out of the home, keeping the inside air cleaner. However, chimneys required the support of a weight-bearing wall, so kitchen fires and kitchens were moved from the centers to the sides of homes. (Even today, kitchens require extensive wall space to house ventilation systems, electrical wires, and pipes for water, waste, and natural gas, which is why they are still typically consigned to the corners of our homes.)

Kitchens came to the New World packed aboard sailing ships. If there was one thing European immigrants brought more of than clothes, it was probably kitchen implements. Second-wave immigrants were told by those who preceded them to bring an "iron pot, kettle, a large frying pan, a gridiron, two skillets, a spit," and wooden dishes.

Then came the Industrial Revolution, which smashed through the kitchen like a steam locomotive. Except for plates, pots, and basic utensils, all the gadgets and gizmos that populate your kitchen emerged from the technological renaissance that began just 200 years ago. Before that time, most of the devices we now take for granted couldn't even have been imagined. The modern kitchen and its labor-saving devices owe their existence to several key industrial-age innovations. Modern plumbing was one. The stove was another.

Cooking actually leaped out of the fireplace and into metal boxes before the official start of the Industrial Revolution, beginning with the invention of a three-sided fire box in 1630. A century passed, however, before a stove was invented that completely enclosed the fire. Another hundred years went by before a British ironworker invented a cast-iron stove in 1802. The new design included a flue for removing smoke from the kitchen, an improvement that accelerated the adoption of stoves. Around the same time, German inventor Frederick Albert Winsor demonstrated a gas-powered stove, but another 24 years went by before James Sharp invented a practical gas stove.

Electric stoves started showing up as soon as Edison began stringing wires to carry electricity into homes. But the earliest electric stoves had design problems and cost a lot of money to power. It wasn't until the 1920s that they became a popular kitchen appliance.

A vital component of the modern electric stove—in fact, of all the electric heating elements in today's kitchens, in everything from toasters to popcorn poppers—is a metal alloy called Chromel or Nichrome, which was invented by Albert Marsh in 1905 (patent number 811,859). He mixed nickel with metals in the chromium group—chromium, molybdenum, tungsten, or uranium—to create an alloy with a high melting point, a low rate of oxidation (formation of rust), and an electrical resistance 50 times that of copper. A high resistance means that when an electric current flows through the material, the moving electrons bump into ions (charged atoms or molecules) in the metal. This increases the energy of those ions, causing them to vibrate, and these vibrations generate heat. Marsh's alloy was used to create the first heating wire that worked reliably without burning out. Nichrome is used to this day to toast bread, pop popcorn, etc.

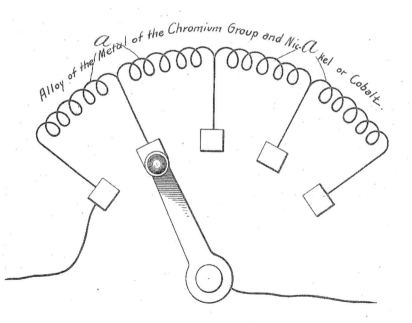

Patent no. 811,859

Another inventor, the offbeat genius Nikola Tesla, paved the way for the modern kitchen when he patented the alternating current (AC) motor in 1889 (patent number 416,194). His invention helped change how electricity was made and distributed. Alternating current gained favor over the direct current system that Edison was advocating, which required a smoke-belching power station every few miles throughout a city. AC power can be transmitted long distances at high voltages, reducing the loss of energy in the wires. All that's required in consumers' backyards is a substation to reduce the voltage and send the power into their homes.

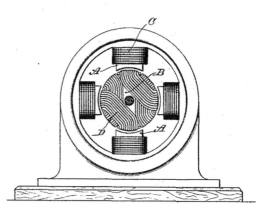

Patent no. 416,194

Think of all the devices in the kitchen that use electric motors. There are some that are direct current (DC) motors—any devices that draw power from rechargeable batteries, such as Dustbusters. But aside from these, any kitchen appliance that plugs into a wall outlet and includes movement uses an AC motor. This includes dishwashers, refrigerators, fans, blenders, coffee grinders, and many more.

The refrigerator, of course, was also a key innovation. Its predecessor, the icebox, allowed people to keep food fresh for several days without salting or drying it, but it required a ready supply of ice to provide the necessary chill.

Just as metals changed the design and use of cutting tools, so too have plastics changed how kitchen appliances are made. Lightweight plastics can be molded into almost any shape and can be produced for a fraction of the cost of other materials. By coupling plastic components with solid-state electronics (and a readily available source of electrical power), designers have created increasingly more complex devices that do almost all the manual tasks people once had to do on their own.

In short, every discovery and invention, from alloys to electricity to refrigeration, ultimately finds its way into the kitchen. The history of kitchen tools and appliances mirrors the history of scientific and engineering innovation.

A Note on Patents

Throughout this book I include references to and illustrations from many of the patents granted to the developers of kitchen tools and appliances. Unless otherwise indicated, the patents listed are U.S. patents registered with the U.S. Patent and Trademark Office. Patents

The Food We Eat

Technical innovations haven't just changed the ways we prepare, cook, and store food; they've also changed how we obtain our food in the first place. Supermarkets, like everything else, had to be invented. Before supermarkets, a grocer, standing behind a counter, would fetch the items on your list for you. It was Clarence Saunders of Memphis, Tennessee, who came up with a store layout that would allow shoppers to do their own fetching and save him the labor costs. This also allowed the store to be bigger (a single grocer wouldn't want to run to the end of Aisle 14 to get your gluten-free organic pasta shells) and therefore to offer a greater variety of products. Saunders opened the first Piggly Wiggly supermarket in 1916, and the concept spread quickly. He was granted patent number 1,242,872 in 1917.

are assigned numbers sequentially; the system was instituted in 1836, with the first patent that year being numbered 1. Earlier patents were given sequential numbers with the suffix "x." So the first U.S. patent, awarded in 1790, was given the number 1x.

Often it isn't clear which patent represents the critical discovery that allowed a new product to come to the public. Patent attorneys can argue for hours (or days, or years) and still not gain clarity. I have selected patents that I believe represent the earliest development of a technology applied to a kitchen task.

If you want to read the relevant patents yourself, a complete database of U.S. patents is available at Google Patents, www.google.com/patents. To find a particular patent, search for the patent number listed in the text or underneath the illustration.

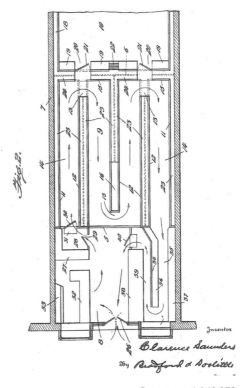

Patent no. 1,242,872

ALUMINUM FOIL

History of Aluminum Foil

Alfred Gautschi of Switzerland invented aluminum foil and was awarded a U.S. patent in 1909 (patent number 917,285). His patent claims the utility of aluminum foil for "packing chocolates and other eatables" and outlines a process for making sheets of aluminum foil that are thinner than $\frac{1}{10}$ of a millimeter. The first use of aluminum foil in the United States was to protect candy, such as Life Savers. It replaced the thicker and more expensive tin foil in American kitchens in 1913.

How Aluminum Foil Works

Aluminum is a metal, and as such it is malleable—you can bend it without damaging it. It also doesn't corrode easily. Unlike iron that rusts in the presence of oxygen, aluminum

Shiny Up or Shiny Down?

Which side of aluminum foil should face the food? Common physics sense suggests that the shiny side should reflect more radiative heat, and therefore should be on the outside for refrigerated food (to keep the heat out) and inside for cooking (to keep the heat in). But in actuality there is very little difference between the amount of reflection from the two sides—certainly not enough to concern the leftovers chef.

is slow to react with oxygen, and when it does, it forms a surface barrier of aluminum oxide that protects the aluminum atoms below it. Kitchen aluminum foil, which is manufactured to a thickness of about 2/10 of a millimeter, protects food by keeping out oxygen, light, bacteria, and water. Other favorable properties of aluminum are its material strength and low weight, and its high heat conductivity—which means that heat passes easily through the foil. Don't wrap your body in aluminum foil before going skiing!

You will notice that one surface of aluminum foil is shiny and the other is dull. This is the result of the manufacturing process. Two sheets of foil are squeezed between rollers and later separated. The inside surfaces of both sheets are dull, but the sides that were facing the rollers are shiny.

BAG SEALER

History of the Bag Sealer

The plastics revolution brought us, among other things, thermoplastics. This family of plastics melt with the application of heat and then refreeze when they cool, making them wonderful materials for creating permanent seals.

Robert Hubbard invented a plastic bag sealer intended for kitchen use. His 1974 patent (patent number 3,847,712) mentions sealing plastic bags for "sandwiches and other food items." His patent was assigned to Dazey Products, the company that made the model that is disassembled in this chapter.

Earlier inventors had discovered a variety of ways to seal thermoplastic bags; one of the earliest is U.S. patent number 3,214,317. However, Hubbard's design appears to be the first intended specifically for home use.

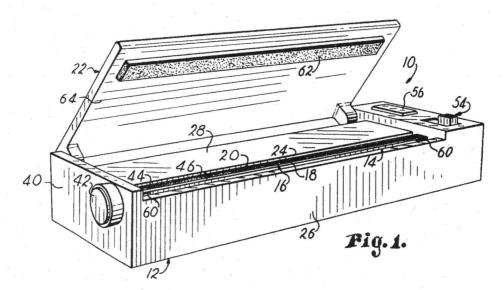

Patent no. 3,847,712

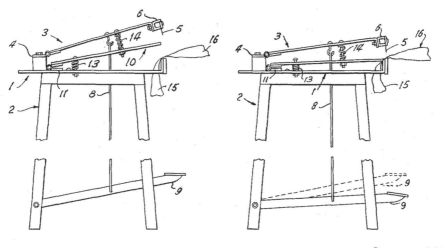

Patent no. 3,214,317

How Bag Sealers Work

The model depicted in this chapter switches on when you lift the lid, but it takes a couple of minutes for the heater to warm up. Then you lay the end of the bag full of leftovers on the sealer's metal edge and close the lid. This compresses the two sides of the mouth of the bag together.

With the lid open, current flows through the heating element, a high resistance wire wrapped in a white insulator, which is under the metal edge. (For more information on how heating elements work, see the introduction, p. xiii.) When you shut the lid, a lever depresses a switch that opens the circuit, cutting power to the heating element. But by this point the edge has grown warm enough to melt the plastic on both sides of the bag, welding them together.

Some more recent models withdraw air from inside the bag to form a vacuum seal. Removing the air from a bag also removes most of the airborne spores and microbes that can spoil food. If you don't have one of these newer models, it's a good idea to squeeze the bag to rid it of air before sealing.

Heater switch

Inside the Bag Sealer

A safety device is wired in series with the heating element under the lid. This small device acts as a thermal fuse. If the temperature rises to an unsafe level, the fuse will open the circuit and stop the flow of electricity. The fuse contains two strips of metal that are joined together with a material that melts above 300° F. Unfortunately, the fuse cannot be reset; if it breaks, you must have it replaced.

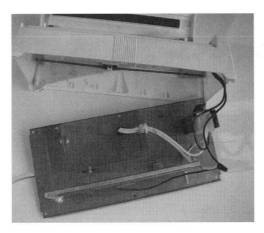

Thermal fuse

BLENDER

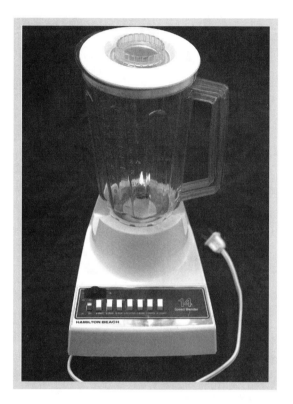

History of the Blender

Stephen J. Poplawski invented the electric mixer in 1922 (patent number 1,480,915) to make malted milk shakes. The Hamilton Beach Company later purchased Poplawski's patents and started making his milk shake blender.

The first blender designed for home use was invented by Fred Osius in 1937 (patent number D104,289). He recruited musician Fred Waring to provide financing and marketing. The product was sold as the Waring Blendor. In 1940, Stephen Poplawski developed his own home mixer (patent number D123,509) and sold it to the John Oster Manufacturing Company. It marketed the new device as the Osterizer starting in 1946.

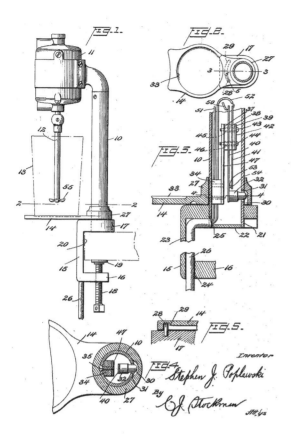

Patent no. 1,480,915

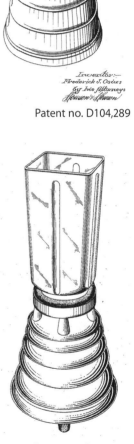

Patent no. D104,289

Patent no. D123,509

How Blenders Work

Electricity is supplied to the base of a blender through a switch that allows you to choose the speed of blending. The switch usually consists of mutually exclusive buttons that you depress to mix, blend, liquefy, etc. You push them and the blender makes a racket as its blades spin. As you push buttons farther to the right, the blades spin increasingly faster. Pushing different buttons changes the flow of electricity to the motor inside; the higher the voltage, the faster the motor spins.

As the motor spins, the blades hurl the contents—let's say, yogurt—toward the outside of the container walls. They force more and more yogurt outward, where it is trapped. The only way it can move is upward, so up the sides of the container the yogurt goes. Along with it go many molecules of air that are drawn into the stirring blades and pushed into the yogurt. The spinning blades pull yogurt in from above and press it out to the sides; this whirling flow of fluid is called a vortex. (Vortexes play important roles not just in blenders but also in sci-fi movies and in your bathtub.)

The powerful motor is the heart of the blender. However, equally important to its operation are the two seals that keep liquids in and out. One seal has to keep liquids in the mixing container on top. Without a reliable seal, you'd have to deal with a puddle under the container. The second seal must keep liquids out of the base of the blender so the motor and electrical controls stay dry. If this seal failed, things could get ugly. The solution is the packing seal known as the O-ring, which was patented by Niels Christensen in 1939 (patent number 2,180,795). During World War II this invention was deemed a war-critical technology, so the U.S. government purchased it from Christensen and made it available to American industry at no cost.

Patent no. 2,180,795

Inside the Blender

At the bottom of the blender's mixing container are the stirring blades. They sit atop a shaft that passes through the container and connects to a rubber or plastic coupler on the underside. Turning this coupler spins the stirrer inside. The trick in making the pitcher is allowing the shaft to turn easily without allowing liquids to leak out. This feat of engineering wizardry is accomplished with an O-ring. This rubber toroid (donut shape) fills the space between parts and expands horizontally as it is compressed vertically. As pressure is exerted on the O-ring, it forms a better seal.

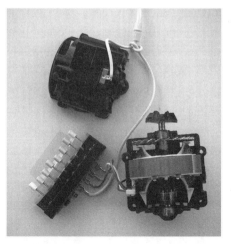

The coupler on the bottom of the mixing container fits into a similar coupler on the top of the base unit. Inside the base, the second coupler is attached to a shaft, which is connected to the blender's motor. As the motor spins, it turns the couplers and thus the mixing blade. It also turns a fan inside the base unit, which draws air past and through openings in the motor to keep it cool. The motor shaft is held in place by metal bearings, one at the bottom of the shaft and one just above the fan.

The base also contains another interesting component: the blender's switches. Eight buttons control a bank of four switches. Depressing a button pushes four sliders to the right or left. These sliders, with angled cuts, lift switch arms as they slide, breaking the circuits. Otherwise, springs keep the switches in contact and the circuits closed.

Across the contacts is a 3.0-amp "general purpose rectifier." This device "rectifies" the alternating current provided through your electrical outlet, converting the cycling positive and negative current into constantly positive current and removing the variations, which provides the blender's motor with steady direct current. This allows the motor to function as a DC motor, which is much easier and less expensive to control the speed of than an AC motor. A DC motor's speed is determined by the voltage applied to it, which is how the blender switches can make the blender spin at several different speeds by increasing or decreasing the voltage.

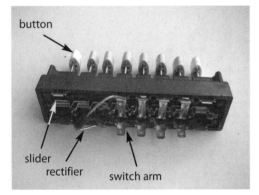
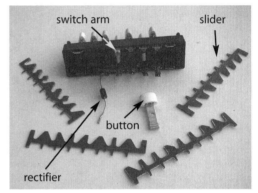

What's the Difference?

What is the difference between a blender, a food processor (p. 69), and a mixer (p. 108)? A mixer usually has one or two detachable beaters that you insert into a bowl of ingredients to mix. With blenders you drop the ingredients, mostly liquids, into an upright container. Blenders give you a wide variety of blending speeds. Food processors, on the other hand, are designed to chop and dice solid food; they operate at one speed with a manual on/off mode. As with a blender, the food processor's container unit sits on top of the motor base, but the container has a center opening for a long motor shaft, and the chopping blades are fit onto the shaft from above.

BREAD MACHINE

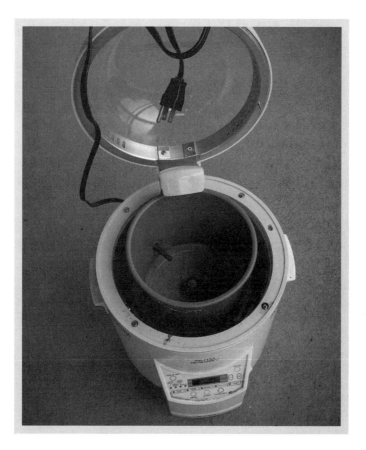

History of the Bread Machine

The first U.S. patent for a bread machine (patent number 383,938) was issued in the late 19th century—before electricity was even available.

The modern home bread machine was invented in Japan and patented in the United States in 1985 for the Hosiden Electronics Company (patent number 4,538,509). The product was a success, which surprised many experts in the kitchen appliance manufacturing field. But once sales took off, many companies developed their own models.

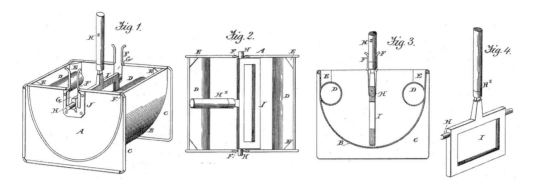

Patent no. 383,938

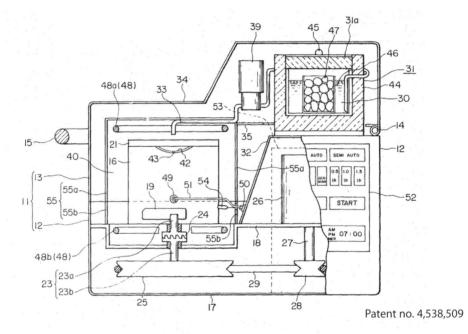

Patent no. 4,538,509

The original home model included a tank so water could be added automatically and an ice box so the water could be cooled to prevent the bread temperature from rising too high. More recent models have done away with these components; thermostats are better able to control the temperature inside the machine.

How Bread Makers Work

It's ingenious that someone thought of integrating all the steps required to make bread—mixing the ingredients, kneading the dough, letting the yeast rise in a warm environment, and then baking the bread—into one tabletop device. One motor, a heating element, a timer, some switches, and a sensor are basically all that's needed to make a bread machine.

Most machines make loaves of either 1.0 or 1.5 pounds. You measure out the ingredients (flour, water, yeast, sugar, a pinch of salt) and drop them into the pan. You set the timer, and the machine takes over. A mixing paddle mixes the ingredients for a certain amount of time. Then the dough is allowed to rise: With the heating element on, the motor stops to give the yeast time to convert sugar into carbon dioxide and alcohol. The flour/water mixture becomes elastic enough to capture the carbon dioxide and form tiny bubbles throughout the bread. Next, the motor kicks on to knead the dough, letting excess gas escape. Finally, the heating element comes on to bake the bread, which also removes most of the alcohol. After the programmed baking time has elapsed, the beeper tells you it's ready.

Inside the Bread Machine

Removing a few screws allows the bread machine's metal outer cover to come off. The front control panel lifts out with its circuit board. Beneath another metal cylinder is the motor and the spindle that the mixing paddle rides on. The motor turns a large plastic geared wheel that turns the spindle above it. A rubber belt connects the motor, which sits off to one side, to the wheel.

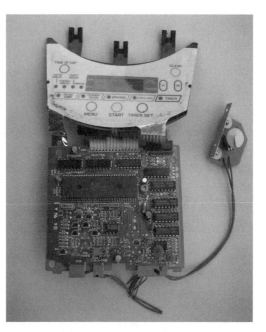

Inside the spindle is a temperature probe called a thermistor. It has a resistor (an electronics component that resists the flow of electricity) with a special property: its resistance changes with the temperature. It sits inside the spindle so it can get an

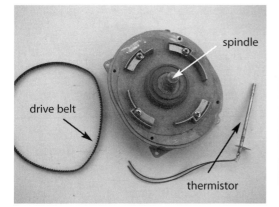

spindle

drive belt

thermistor

Heating element

accurate reading of the temperature inside the baking loaf of bread and signal when the heating element should turn on and off. The heating element itself is a coil of high resistance wire. (For more information on how heating elements work, see the introduction, p. xiii.)

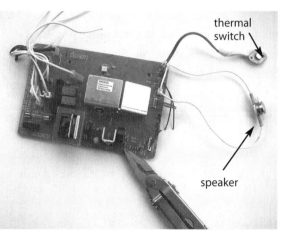

thermal switch

speaker

A circuit board located near the motor connects to two thermal switches. They monitor the temperature of the machine's inner

centrifugal fan

metal cylinder to ensure that it doesn't overheat. The circuit board also includes the piezo speaker that bleeps at you when the bread has finished baking. (A piezo speaker contains a crystal that vibrates and makes sound when it receives a changing electrical voltage.)

At the bottom of the bread maker is a second motor. It directly drives a centrifugal fan that draws air in from beneath the bread maker and pushes it out between the machine's inner cylinder and outer cover. This is one more precaution designed to prevent the user from getting burned. Not all machines have this second motor.

Bread machines seem to do best with wheat flour—flour that contains gluten.

CAN OPENER

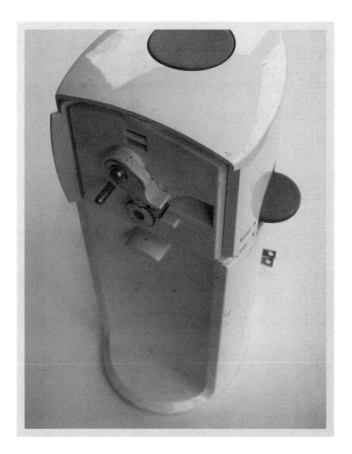

History of the Can Opener

Consider this curious historical fact: tin cans were in use for nearly half a century *before* the can opener was invented. Based on the existing technique of preserving food in glass bottles, Peter Durand invented tin cans in 1810. The first U.S. patent for tin cans was awarded in 1825 to Thomas Kensett. The cans used then were made of thick metal. By mid-century metallurgy had improved so cans could be made of more lightweight metals. At this point, inventors turned their attention to how to better open these cans.

Does that mean that canned goods sat on pantry shelves for 45 years before people figured out how to open them? No, people were ingenious enough to open cans without dedicated openers—chisels, strong knives, or other tools were jammed into the cans to get to the food. No opener? No problem! Set the can on a rock in the fire and stand back. Eventually the can explodes, spewing beans all about and leaving half a can heated and ready to eat. (I witnessed this myself on a Boy Scout camping trip.)

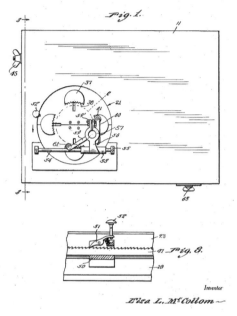

Many inventors turned their attention to can openers in the mid- to late 19th century. Ezra Warner invented the first can opener in 1858 (patent number 19,063). The type of opener used today, employing a cutting wheel, wasn't invented until 1870, 60 years after canned foods were introduced. The first electric can opener was created by E. L. McCollom in 1932 (patent number 1,892,582).

Patent no. 1,892,582

How Can Openers Work

Is a lid of steel standing between you and your dinner of baked beans? Lift the can so the cutting wheel of the can opener is positioned against the inside rim of the can. Push down on the lever to drive the cutting wheel through the metal lid. Then, by hand (for nonmotorized models) or motor, rotate the can so its entire circumference passes under the cutting wheel.

The electric can opener has an electric motor that starts turning when the can opener's lever arm is pushed down. The motor rotates a serrated wheel, which in turn rotates the can under the cutting edge. Many can openers employ a magnet supported by a spring arm to grab the lid and hold it after it has been cut free from the can. Some have a knife sharpener on the back; the same motor that powers the opener turns a grinding wheel that can be used to sharpen knives.

Inside the Electric Can Opener

An electric motor constitutes about half the weight of the entire machine. On the motor shaft is a small pinion gear with teeth that are cut at an angle to the shaft. This is a spiral gear, which transmits high-speed motion with little noise or vibration. The spiral gear drives a much larger plastic gear, which shares its shaft with another small gear. That gear drives an even larger gear made of metal. This arrangement of small gears driving larger ones reduces the speed of rotation and increases the torque, or turning power. The electric motor spins way too fast to be useful without gearing, so the gears slow it down and help it deliver more turning "oomph."

The only other items of interest are the switch and cutting blade. The one-sided blade is

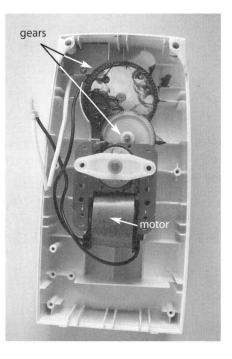

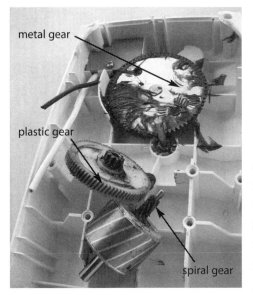

common among can openers. It punctures the metal lid of the can while the motor rotates the can beneath it. The can opener's power switch is a simple affair. You depress the plastic button on the outside and it moves the metal arm down to the contact, which completes the circuit and energizes the motor.

Why Tin Cans?

Do you call them "tin cans"? Food cans were originally made of tin with lids soldered on, but today's canned goods come in containers made from tin-plated steel (the tin provides rust protection for the steel) or, in the case of more lightweight cans, from aluminum.

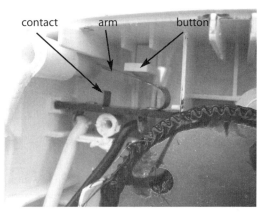

COFFEE GRINDER

History of the Coffee Grinder

Hand-turned coffee mills have been in service for centuries. The electric grinder was invented in 1907 (patent number 856,167). The inventors envisioned that it would be used for grinding both coffee and spices.

How Coffee Grinders Work

Milling or grinding coffee beans greatly increases the surface area of the coffee so the hot water can extract more of the oils that make a good, strong cup. If the milling is too coarse, the surface area of the beans will be small (relative to the volume of beans), and the low amount of oil released will make a weaker cup. But if the milling is overly fine, too much of the bean's surface will be exposed to the hot water, releasing too much oil and making the coffee taste bitter. So the job of a grinder is to break up the beans to the optimal size.

Hand-cranked mills use rotating grinders to crush the beans. Electric versions of mills (unlike grinders) spin metal blades at low speeds of 500 RPM.

Grinders run at much higher speeds; the one shown here operates at about 20,000 RPM. It cuts the beans with a two-armed blade. This design is found most often in home kitchens; it is inexpensive and works forever. However, the size of the grounds can vary greatly with this style of grinder, decreasing the quality of the resulting coffee. Burr grinders and roller grinders, which don't use the fast-spinning two-armed blades, provide more uniform grinds but are more expensive.

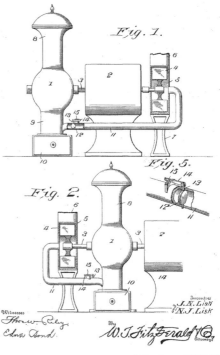

Patent no. 856,167

Inside the Coffee Grinder

Two Phillips screws hold the motor assembly in the plastic body. With some ungentle coaxing, the motor comes out. The blade is screwed onto the motor shaft. Interestingly, the motor shaft is gimbaled, allowing the blade to move up and down as it spins around, allow-

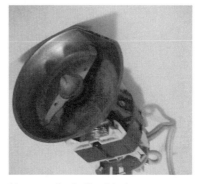

Motor and grinding blade

ing it to reach all the beans. The motor is sturdy and probably outlasts the rest of the grinder.

The switch to power the motor is at the base of the grinder. To operate the grinder, you align a plastic tab on the lid with a slot on the base. Pushing the tab in place depresses a plastic arm that stretches to the bottom of the grinder, where it slides metal contacts. A plastic bow acts as a spring to push the arm up and cut power to the motor when you remove the lid.

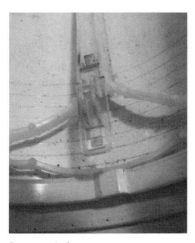

Power switch

COFFEE MAKER

History of the Coffee Maker

The story of coffee's discovery—perhaps just a myth—is that a goat herder in Ethiopia saw goats eating coffee beans and noticed that it gave the animals unusual vitality, so he tried the beans himself. From the hills of Ethiopia to a Starbucks on every corner—coffee is more than a drink; it is a worldwide phenomenon.

Throughout most of the history of coffee, from its discovery in the 9th century into the late 19th century, people prepared this beverage without coffee makers—they just tossed a handful of grounds into a pot of boiling water. In many places throughout the world, people still use this technique. Once their coffee is brewed, some of them pour it through a filter to separate out the grounds; others just pour it into a cup and let the grounds settle to the bottom.

The vacuum coffee maker was invented in Berlin in 1830. It consists of two nearly identical containers, usually made of glass, one sitting on top of the other. Coffee grounds are put into the upper container and water into the lower one. When the device is set on the stove, the water heats up and is forced by pressure into the upper bowl. When it's removed from the stove, the lower bowl cools and creates a partial vacuum, drawing the brewed coffee back down through a filter. This is the coffee maker I remember from my grandmother's kitchen. As a six-year-old watching her and my grandfather fix breakfast, I was mystified by this magical device.

American inventors started to create their own versions in the mid-1800s. In 1860, Thomas Yates invented one of the earliest American coffee makers based on the vacuum design (patent number 28,803).

A different type of coffee maker, the percolator, appeared in American homes long before electricity. James Nason invented the first one in the United States at the end of the

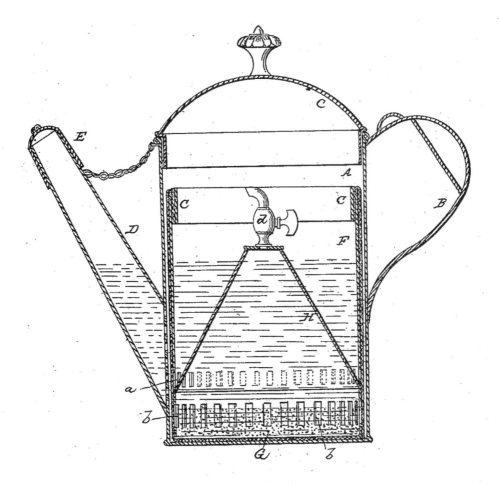

Patent no. 28,803

Civil War, in 1865 (patent number 51,741). Percolators repeatedly send water through the grounds and back into the pot. A well in the bottom of the percolator pot collects water or coffee, heats it to a boil, and sends it up the hollow tube. It spurts out the end, hitting the viewing cap, a clear plastic or glass window into the dark world of the percolator, then drains down through the coffee grounds before falling back into the pot below. When electricity became available in the early 20th century, electric percolators became the coffee maker of choice.

In 1971, Vincent Marotta invented the drip coffee maker known as Mr. Coffee (patent number D229,158), and he and a partner started a company to manufacture them. The next year the company introduced the product to consumers, with an adver-

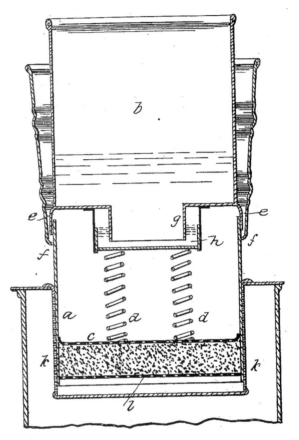

Patent no. 51,741

tising endorsement from baseball's Joe DiMaggio, and drip coffee makers quickly displaced percolators in American homes. The difference in taste between percolator-made coffee and coffee made by drip is due to the fact that in the percolator the coffee must be kept at boiling temperatures even after it's brewed, which can rob the drink of some of its flavor.

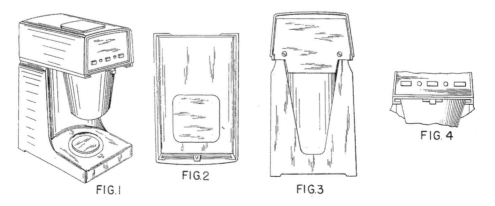

FIG.1   FIG.2   FIG.3   FIG. 4

Patent no. D229,158

How Coffee Makers Work

Passing hot water over ground coffee beans releases a host of chemicals—over 500 compounds—including caffeine. Each type of coffee machine performs the task a bit differently.

The machine shown here is a drip maker. To use it, you pour fresh water into a reservoir and add ground coffee beans to a container on top of the coffee pot. Turning on the power allows electricity to pass through the heating element that curves around the inside of the base, below the coffee pot. The heating element touches the tube through which the water passes. This heats the water in the tube, eventually bringing it to the boiling point. Resulting bubbles rise up the vertical tube, pushing hot water ahead into the container with the ground beans. You can hear the burping sound of bubbles of steam and water rising up the tube. The hot water drips down onto the ground beans and collects in the pot below. A filter keeps out the grounds while letting the liquid pass.

Older methods of brewing coffee include the vacuum and the percolator, discussed in the "History" section, above. Another method still in use today is the coffee press, in which coffee grounds are combined with boiling water in a pot and later filtered out when a sieve is pressed down through the container. Espresso is made by using pressure to force hot water through the grounds; see the full discussion of the espresso maker on p. 57.

Inside the Drip Coffee Maker

The top of the coffee maker pops off if you use a screwdriver as a wedge and a small saw to cut through the plastic rivets. Inside is the reservoir that holds the water. There is a drain in the bottom so the water you add can flow into the heating tube below. Next to the drain is a plastic standpipe that brings up the hot water after it has passed through the heating tube. The water rides up in a rubber hose inserted into the standpipe. At the top of the standpipe is a plastic tube that carries the hot water over to the filter. There the water squirts out and drips into the filter below.

The bottom of the coffee maker pries off to reveal the heating tube. Really it is two tubes, one on top of the other. The bottom tube carries the water and the top one holds

Top view

the heating wire. (For more information on how heating elements work, see the introduction, p. xiii.) The heating tube is on top so it can both heat the water to make coffee and then keep the pot of brewed coffee warm.

The power cord carries electricity into the machine. One of the two wires passes through the on/off switch. From there the electricity goes

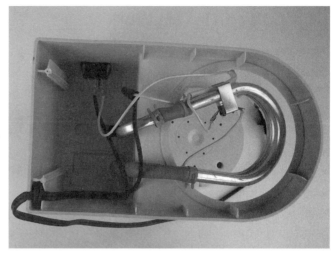

Bottom view

through two thermal devices. The first is a thermal fuse. If something malfunctions and the temperature inside the coffee maker rises to dangerous levels, this one-shot device will interrupt the electric current. It will not reset; if blown, it must be replaced.

The second device is a thermostat that turns the power to the heating element on and off. This is what produces the "ka-blink" sound you hear as the heating coil cycles on and off. The device is a bimetallic switch that has parts that expand and separate when heated, breaking the circuit. When they cool they contract, come together, and close the circuit so the heating element receives electricity.

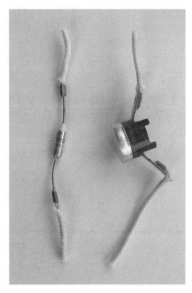

Thermal fuse (left) and thermostat (right)

Inside the thermostat

COFFEE ROASTER

History of the Coffee Roaster

Roasting gives coffee beans their characteristic flavor and smell. Once roasted, the beans lose their flavor over time, so some people prefer to roast their own coffee at home.

Some early patents for coffee roasters were granted in the late 19th century. Patent number 204,067 was issued in 1878; number 268,724, in 1882.

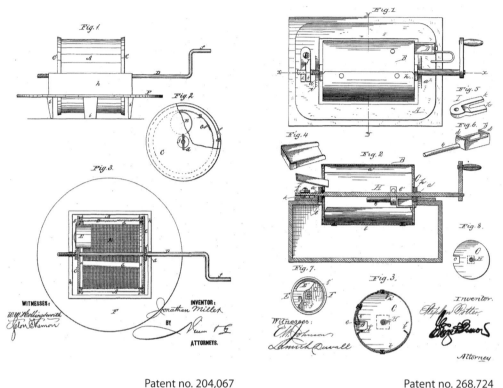

Patent no. 204,067

Patent no. 268,724

How Coffee Roasters Work

Temperature and duration of roasting determine the flavor coffee beans can produce. The roasting process can take from a few minutes to about half an hour. (The machine pictured at the beginning of the chapter roasts a pound of coffee in 20 to 30 minutes.) Beans are roasted at temperatures ranging from 400° to 480° F. At about 375° F, the beans crack, producing an audible sound, like popcorn. With continued heating the beans crack again, which signals that the beans are at a medium roast. Roasting beyond this point produces darker roasts. The roasting process also produces a strong aroma from the beans' oils burning, so it is best to do it outside.

Roasting machines not only heat but also stir the beans to ensure that they are roasted uniformly. Some, like the one depicted at the beginning of the chapter, use a vertical auger to continuously pull the beans up and away from the heating element, allowing other beans to fall toward the heater. Others, like the one disassembled in the next section, use blowers to keep the beans moving.

Some coffee lovers show great ingenuity by creating homemade coffee roasters out of other kitchen appliances, such as bread machines and popcorn poppers. A popcorn popper makes an effective and inexpensive roaster. Some tinkering is required to place the heating element and the fan on separate controls. The heater is turned on and the fan is adjusted via dimmer switch during the roasting. This allows the roaster to blow out the beans when their color indicates that they are roasted to his or her preference.

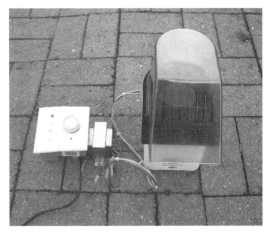

Homemade coffee roaster

Inside the Coffee Roaster

Removing the plastic cover reveals the metal cylinder where the beans are roasted. Beneath the cylinder is the heating element, a coil of high resistance wire. (For more information on how heating elements work, see the introduction, p. xiii.) The flow of electricity to the wire is controlled by a flip switch. In this model, to keep the beans from burning, a blower in the base beneath the heater blows air upwards through the pile of beans. The blower is controlled by a separate circuit that the user operates with a dimmer switch–like control. A heating element, a blower or other stirring device, and their switches are all that make up a roaster.

COFFEE URN

History of the Coffee Urn

Many inventors have contributed ideas to the technology of making coffee. One of the earliest electric coffee urns was patented by C. C. Armstrong in 1931 (patent number 1,816,994). His invention included an automatic heat cut-off after enough water had passed through the grounds, rendering the coffee-making process automatic.

How Coffee Urns Work

The coffee urn is a large percolator. (See the chapter on coffee makers, p. 20, for more information.) It sends hot water up a tube to splash onto the grounds and drip back into

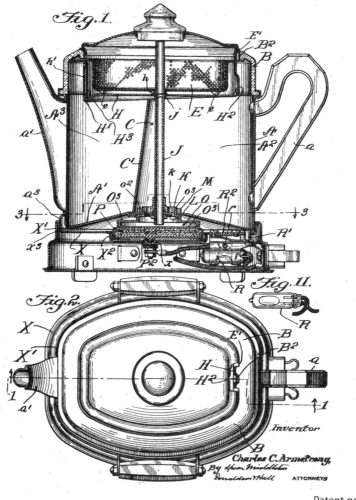

Inventor
Charles C. Armstrong,
By Man Middleton
Donaldson Hall   ATTORNEYS

Patent no. 1,816,994

the reservoir. When the water/coffee in the reservoir has been heated to the right temperature, the thermostat cuts power to the heating coil and turns on the ready light, telling you to grab a cup.

Inside the Coffee Urn

Everything interesting is in the base. The top is just a reservoir for holding water before brewing and holding coffee afterward.

One screw holds the black plastic cover over the heating component in the base. The electrical wires lead to a thermostat—the thing that makes that "pop" sound when the pot turns on and off—and a small light. The light is the "ready" light. It is wired in parallel to the thermostat. When the urn is plugged in, current flows through the open thermal switch to the heating coil, starting the coffee brewing. When the urn is filled with hot coffee instead of cold water, the thermal switch turns off. This forces electricity through the light instead, announcing that coffee is ready. But in this mode, current still flows through a warming coil to keep the coffee hot.

The heating element is embedded in the side of a small well that protrudes from the center of the reservoir's base. Inside the reservoir, the water that fills the well is heated, and then it bubbles up the central "stem," which also sits in the well, toward the basket filled with coffee grounds. The heated water squirts out the top of the stem and over the grounds.

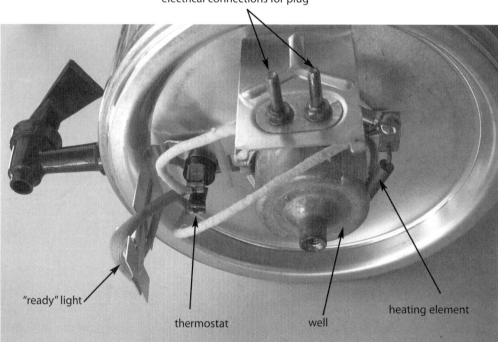

electrical connections for plug

"ready" light

thermostat   well

heating element

CORK REMOVER

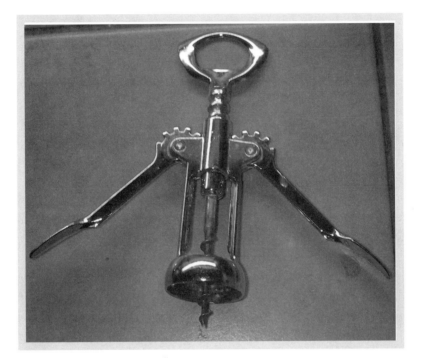

History of the Cork Remover

Although people have been making and drinking wine for 8,000 years, only recently did we begin storing wine in glass bottles for which we need devices to remove their corks. In ancient times, wine was stored for only short periods of time, in wooden barrels or pottery vessels, before it was drunk. As glass-bottle-making technology improved in the 18th century, sealable bottles were invented. The English stoppered their bottles with cork from the bark of trees found in Portugal and Spain.

The first corkscrew was invented in the late 18th century in England. The first U.S. patent to mention removing corks from a bottle was granted in 1856. Four years later, M. L. Byrn was awarded a patent for a device he identified as a corkscrew (patent number 27,615).

Patent no. 27,615

How Cork Removers Work

Most corkscrews give you some mechanical leverage to pull the cork up, which is helpful, because it takes between 25 and 100 pounds of force to yank out a cork.

The model shown at the beginning of the chapter, called a wing corkscrew, is a popular, easy-to-use design. You twist the top handle, which doubles as a bottle cap remover, to embed the screw in the cork. As the screw buries itself in the cork, the wing handles rise out to the sides. By pushing down on these handles you use their leverage to pull the cork up and out.

Simpler is the corkscrew on my Swiss Army knife or the slightly more useable "waiter's friend." Twist the screw into the cork and use lever action to pry the cork out. Either requires a bit of strength. But you can leverage the knife body against the lip of the bottle to budge the cork.

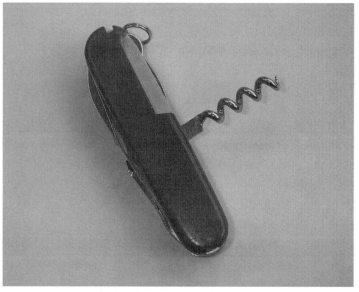

Swiss Army knife corkscrew

Putting It Back In

Once the cork is out, how do you get it back in? Corks expand in size soon after leaving the neck of the bottle, making it difficult to reinsert them into the bottle. Air-pump systems use a rubber cork substitute and a pump. With the rubber stopper in place, each stroke of the pump removes some of the air in the bottle. This creates a partial vacuum that holds the stopper in, and with most of the air removed, the wine will oxidize very slowly. See the chapter on wine savers, p. 198.

Other devices to remove corks include the twin-prong cork remover. You slide two prongs into the bottle, along the outside of the cork. With a twist and pull, the cork slides out. With air-pressure cork removers, you push a needle through the cork to pump up the pressure inside the bottle. As the pressure builds, it forces the cork out of the neck.

Inside the Wing Corkscrew Cork Remover

Everything is exposed; there is nothing to take apart. In a close-up of the wing corkscrew you can see how the gear at the end of each wing meshes with threads on the shaft that holds the screw embedded in the cork. The wings allow you to push down at a distance from the gear, giving you a large mechanical advantage.

CROCK-POT

History of the Crock-Pot

This "thermostatically controlled cooking apparatus" designed for slow cooking and deep fat frying appeared in 1968 (patent number 3,508,485). A second model, patented in 1975 (patent number 3,908,111), included a glass or ceramic insert that could be removed for cleaning. This became the Crock-Pot, which passed through two corporations and became a brand name owned by Sunbeam.

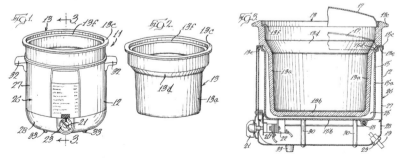

Patent no. 3,908,111

The patent claimed on the bottom of the device disassembled in this chapter is number 6,498,323, assigned to the Holmes Group in 2002.

How Crock-Pots Work

Food is placed in a Crock-Pot along with water or another liquid, and the temperature dial or switch is turned on. The pot maintains the temperature in the range of 175° to 200° F, but can also warm food at a lower temperature of 160° to 165° F. The lid must remain on the pot to keep the food from drying and prevent bacteria from growing.

The base of the cooking pot has a heating element made of loops of highly resistive wire. (For more information on how heating elements work, see the introduction, p. xiii.) The heating element warms the pot and its contents, and a thermostat turns the heating element on and off to keep the temperature in the desired range.

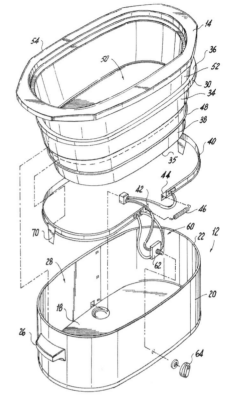

Patent no. 6,498,323

Inside the Crock-Pot

A single nut and screw eye hold this inexpensive Crock-Pot together. The metal base falls away when the nut is removed to show a very simple device. The electrical power cord delivers energy to a coil of heating wire, which is wrapped around the side of the cooking vessel. The vessel itself is made of ceramic, which is slow to warm up, so it allows the Crock-Pot's contents to retain the heat imparted by the electric coil. At the bottom of the ceramic vessel is a metal bar, to which the screw eye is anchored. The screw eye projects through the metal base and is held by the nut.

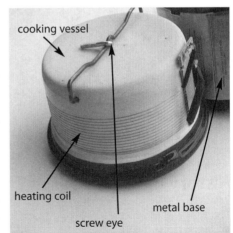

cooking vessel

heating coil

metal base

screw eye

The heating wire, which is coated with lacquer, has a total resistance of 430 ohms. (How do I know that arcane fact? I measured it with a voltmeter.) That means that the electric current flow is about a quarter of an amp, a reasonably small current. There are no other components; this inexpensive model even lacks a thermostat.

Deep Fat Fryer

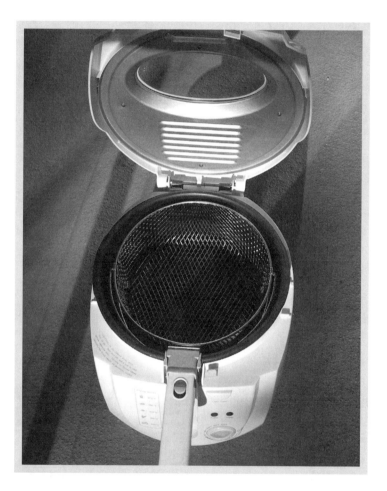

History of the Deep Fat Fryer

The method of cooking food in hot oil is at least 4,500 years old, but it wasn't until 1949 that Stanley Budlane and Robert Dusek invented an electric deep fat fryer (patent number 2,593,392) for home use. They claimed that their design eliminated most of the "excessive smoking" generated in cooking. Their device was thermostat controlled.

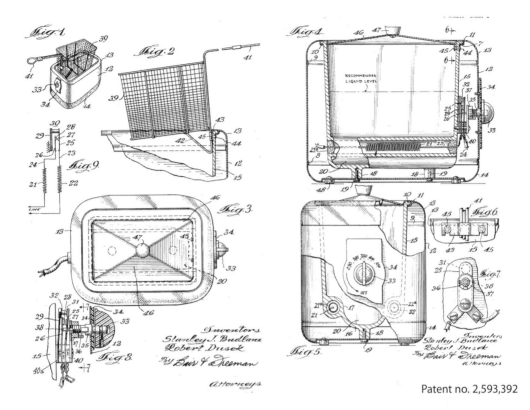

Patent no. 2,593,392

Deep fat fryers didn't catch on in home kitchens until 1976, when the FryBaby electric deep fryer was introduced. It featured a thermostat-controlled heating element and a snap-on lid so the oil could be stored in the fryer without going rancid from oxygen exposure. The following year the larger FryDaddy model hit the market; an even larger model, GranPappy, was introduced a year later.

How Deep Fat Fryers Work

Dropping a slice of potato into the hot oil of a deep fat fryer starts a series of changes that end with a crispy french fry. As the fry warms up, the water inside expands and eventually vaporizes. This water vapor leaves the potato along one of the cut edges and sizzles as it hits the hot oil. After a large portion of the water has been driven off, the edge turns dark and crispy. Cooking continues on the interior of the slice, but much of the water that remains is trapped by the crusty edge so it can't escape.

Why cook the food in hot oil instead of just boiling water? Because fats or oils have a much higher boiling point than water, which means that foods can be cooked at hotter temperatures without the liquid evaporating. Frying foods in oil is quicker than boiling

them, and the food turns out crispy. Frying cooks the outer surfaces of foods and
caramelizes the sugars inside.

Most fryers are heated by an electric heating coil of high resistance wire.
(For more information on how heating
elements work, see the introduction,
p. xiii.) But some more expensive fryers,
including some commercial fryers, use
infrared heaters that transfer electrical
energy into light in the infrared wave-
lengths. Imagine an incandescent light-
bulb that emits no visible light—just
heat. The fryer's bulb has a covering
similar to the glass covering on a light-

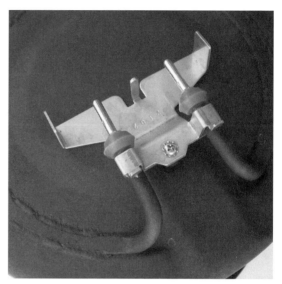

Electric heating coil

bulb. The heat the bulb generates is transmitted by radiation—just like the sun's warming rays. The process is very efficient. Other commercial fryers use gas for heat.

Inside the Deep Fat Fryer

The first interesting thing about this fryer is its electrical connection. It is held in place by
a strong magnet so the power cord will simply fall off if someone kicks it. This kind of
breakaway connector is especially important with a deep fat fryer; if someone tripped on
an undetachable cord and knocked the fryer off
kilter, skin-burning grease could fly everywhere.
In the photo of the connector, the two obvious
metal strips are not what carries the electric current—those are the magnet contacts to hold the
cord onto the cooker. Two pins fit into the holes
on either side of the metal contacts, completing
the electrical circuit but still allowing the cord
to break away if yanked.

Magnetic breakaway connector

The fryer's lid has a vent for steam and oil
vapors. To prevent some of the volatile oils from
escaping into the kitchen air, the lid has two different kinds of filters.

The cooking pot pulls out. Writing on the
bottom tells us that this device uses 150 watts of

power. That's over 12 amps of current! (Most of the circuits in a home are wired so that the circuit breaker will switch off if the current exceeds *20* amps. If another appliance is drawing current from the same circuit as the fryer, the power might shut off.) The plug is stamped with its rating: 15 amps.

Two tiny lightbulbs on the inside tell you when the power is on and when the oil is hot and ready for cooking. The temperature dial rotates a shaft that moves a piece of metal in and out: this is a bimetallic thermostat, with electrical contacts that expand and separate when heated, breaking the circuit. The end opposite to the dial is a copper plug for detecting the temperature. It is a large piece of copper—large enough that its temperature doesn't fluctuate too quickly; copper conducts heat well, so it will quickly pick up the heat from the pot. The copper plug is pressed against the pot with a spring to ensure that it is in constant physical and thermal contact.

Looking closely at the thermostat, you can see the two electrical contacts. Turning the dial toward lower temperatures moves the strips away from each other. They will break contact at a lower temperature. Turning the dial toward higher temperatures makes it harder for the two strips to separate, requiring higher temperatures to shut off the circuit. The thermostat is manufactured by a company in China, and their marketing information says that the main applications for their thermostats are in waffle makers, electric irons, ovens, and rice cookers. This thermostat's temperature range is 265° to 375° F.

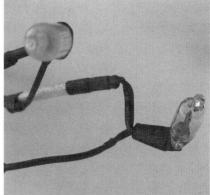

Indicator bulbs

thermostat dial

electrical contacts

DISHWASHER

History of the Dishwasher

Joel Houghton invented the first machine to wash dishes in 1850 (patent number 7,365). This hand-cranked device was designed to clean "crockery or other articles of table furniture." But I couldn't find a record of anyone ever using his device.

Improvements in the dishwasher's design were suggested by many people—including quite a few women inventors. It's rare to find early patents assigned to women, partly because society often discouraged them from pursuing anything technical, and partly for legal reasons—married women didn't have equal property rights, and patents are considered a form

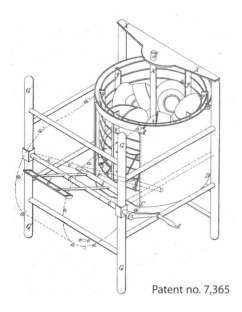

Patent no. 7,365

of property. But women are surprisingly well represented within the patents for dishwashers (or maybe it's not so surprising). Josephine Cochran is generally credited with inventing the practical dishwasher, in 1885 (patent number 355,139), but there are more than 30 other dishwasher-related patents assigned to women from this same time period.

As for Cochran, she rarely washed dishes herself; she employed servants to do that. However, their tendency to chip her fine china while washing it led her to invent the mechanical washer. Cochran was a relative of the inventor of the steamboat, John Fitch.

Homemakers were slow to embrace dishwashers, and a marketing survey in 1915 found that of all the household chores, washing dishes was considered both an enjoyable and a relaxing task. Not until the postwar boom of the 1950s did dishwashers become popular in homes.

Advances keep coming. One I love is the two-drawer dishwasher. You don't ever need to put dishes in the cabinet. You pull clean ones out of one of the two identical dishwashing drawers and put the

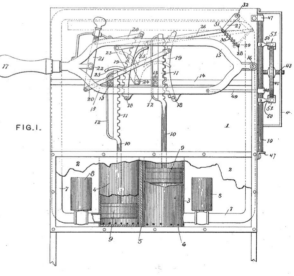

FIG.I.

Patent no. 355,139

Two-drawer dishwasher

dirty ones into the other! The drawback is that each one is less than half the size of a regular dishwasher, so the big things that need cleaning might not fit in.

How Dishwashers Work

If you squirt hot water hard enough and load it with enough caustic chemicals (detergents), you'll get dishes clean. In a dishwasher the water is hotter (130° to 150° F) and the detergents are stronger than you could stand if you washed the same dishes by hand. Simply put, a dishwasher is a waterproof container that heats and squirts water, releases chemicals at the right time, drains the water, and dries the dishes.

An electromechanical timer, just like the one used in clothes washers, controls older models. Newer models come with a microprocessor that allows you to select from a greater range of washing options.

water intake

electrical valve controls

The water pressure from your plumbing system pushes water into the dishwasher, but once it's inside a pump shoots it around to wash the dishes. The pump also drains the water out between wash and rinse cycles and at the end of the rinse cycle. Some dishwashers have *two* pumps, one for spraying the water and one for draining it; both are run by the same motor. Other dishwashers have one pump and an electrical switch that either opens a valve to let pressurized water squirt into the wash tub or closes that valve and opens another to let the water escape into the drain.

Inside the Dishwasher

The dishwasher is a metal shell inside a separate metal housing. The inner shell is waterproof. This is where you put the dishes—where, after you close the door and push the buttons, the work begins. Outside the inner shell, at the lowest point of the dishwasher, is the electric motor. The motor turns the pump. Water that is pressurized by the pump flows into the two spray arms. They turn like lawn sprinklers to spray water throughout the inner shell.

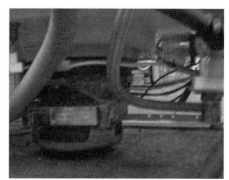

Dishwasher motor

Hand Wash Those Knives

As great as dishwashers are, don't use them to wash your good kitchen knives. Dishwashers can damage knives three ways. First, over time they can destroy the handles, especially wooden handles. Second, the heat inside a dishwasher can weaken the steel. Finally and most destructively, when a knife jostles against its neighbors in the dishwasher's utensil bin, its blades are dulled.

The other component of interest in this area is the solenoid. This is an electrically operated valve that lets water drain out. At the heart of the solenoid is an electromagnet, so when a control circuit sends an electric current to it, the magnet forces the valve open or closed. A float valve on the floor of the inner shell protects your kitchen floor from overflows. The float rises with the water level inside the shell. If the water level rises too high, the float trips an attached switch and shuts off the incoming water.

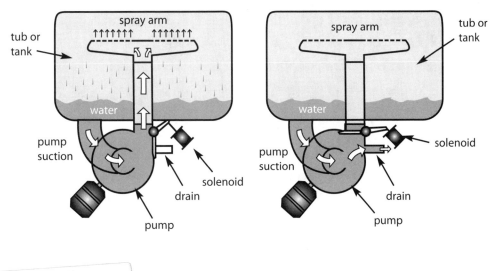

DUSTBUSTER

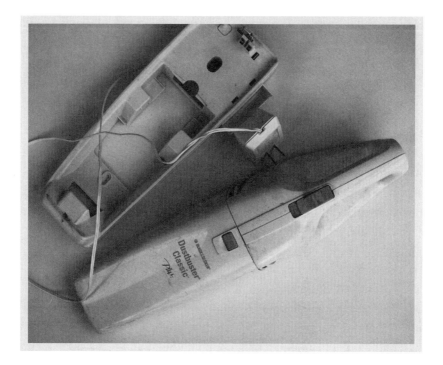

History of the Dustbuster

Designed by Carroll Gantz and patented in 1980 (patent number 4,225,814), the cordless handheld vacuum was first brought to the market by Black & Decker, as the Dustbuster. Prior to the Dustbuster's introduction, handheld vacuum cleaners were powered through an electrical cord plugged into an outlet. The new cordless design was obviously a winner— the Dustbuster sold over a million units in its first year, becoming the most successful product in the history of Black & Decker. Some estimate that 150 million cordless vacuum cleaners, many of them made by Black & Decker, have been purchased worldwide.

How Dustbusters Work

A Dustbuster by any other name is a handheld cordless vacuum cleaner. Vacuum cleaners create low air pressure inside themselves, usually by forcing air out their backside via a motor connected to a fan. Higher ambient air pressure outside of the device forces air into the throat of the cleaner, along with dust, dog hairs, and yesterday's crumbs. The air draws the dirt along with it into a porous bag that arrests the dirt while letting air pass through. The bag can be removed from the machine and its contents shaken into a trash can.

Cordless vacuum cleaners operate on small direct current motors. The motors are powered by batteries, which are recharged when the vacuum is docked with its base and the base is plugged into an outlet. A common problem with Dustbusters is making solid contact with the recharger base. If your unit loses power after a few seconds of operation, either the batteries need to be replaced or they aren't getting recharged.

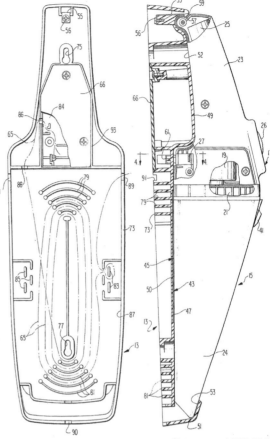

Patent no. 4,225,814

Inside the Dustbuster

The plastic body of a handheld vacuum cleaner has two mirror-image parts held together by a few screws. Removing the screws gives access to the motor, fan, and what little circuitry it has.

If you are doing this at home, first remove the batteries. They are rechargeable but often the first thing to fail. If after years of dependable service, your Dustbuster isn't working, check the batteries with a voltmeter to see if they have a charge. To extract the batteries, open the battery compartment on the bottom of the Dustbuster, usually with a push of the thumbs. The batteries are wrapped in plastic packaging that holds four nickel-cadmium cells.

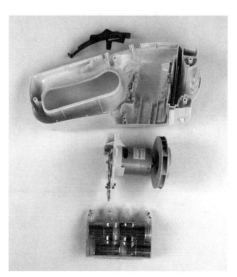

Back to opening the plastic body. Plastic is a wonderful material that allows devices such as this to be manufactured inexpensively. Each half of the chassis is molded plastic. At the factory the parts pop out of the mold, the other components are inserted, and the two halves are screwed together. Compare that process to the one required to manufacture a handheld vacuum from 50 years ago, with metal parts and lots of connectors that had to be assembled by hand. No wonder they didn't catch on then.

Battery pack

The on/off switch is another piece of plastic that when pushed forward forces a piece of copper to make contact with two terminals and complete the electrical circuit. The bend in the copper springs it away from the contacts when the switch is turned to "off."

Copper strips, supported by a piece of plastic, carry the electrical power to the motor. A centrifugal fan is fixed on the motor shaft. As the fan blade spins, it captures air and pushes it outward, creating the low pressure region in front of the motor that sucks in air and dirt.

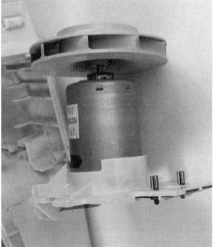

Motor and fan

The recharger unit contains a transformer that changes the voltage from the wall outlet (120 volts) into the voltage that the batteries need (11.5 volts for this model; for others it may be as much as 24 volts). The alternating current then feeds a circuit inside the Dustbuster that includes a diode. The diode passes current in one direction only, so it's an inexpensive way to rectify the alternating current into direct current suitable for charging the batteries.

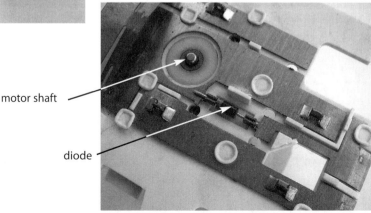

motor shaft

diode

ELECTRIC GRILL

History of the Electric Grill

In 1997, Michael Boehm and Robert Johnson were awarded a patent for the electric grill depicted in this chapter (patent number 5,606,905); their grill was brought to market two years before the patent was issued. And no, former heavyweight boxing champion George Foreman isn't mentioned in the patent, though he is the one person most commonly associated with this grill. He signed on to promote it and he has done so very successfully—some 80 million George Foreman Grills have been sold.

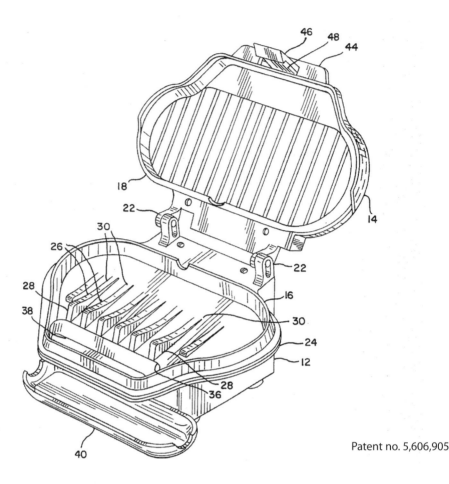

Patent no. 5,606,905

How Electric Grills Work

Key to the success of the electric grill is its nonstick Teflon surface—see the chapter on the Teflon-coated frying pan (p. 159) for more information. Another popular feature is its fat-draining design: The meat, sandwich, or other grillable item sits on raised ridges. Fat that cooks out oozes into the valleys between ridges, away from the food. Then the downward slant of the grill allows the grease to drain into a separate plastic trough that sits underneath the grill.

Plugging the electric grill into an outlet sends electricity through the heating elements in both the upper and lower grill sections. (For more information on how heating elements work, see the introduction, p. xiii.) Electricity also flows to the indicator light in the top to show that the heating elements are receiving power. The machine is rated to use 760 watts of power (about 50 watts less than I calculate that it actually uses).

When the temperature of the food has risen to the preset temperature, the thermostat opens the circuit to stop the electric current. The light goes out, telling you that lunch is ready.

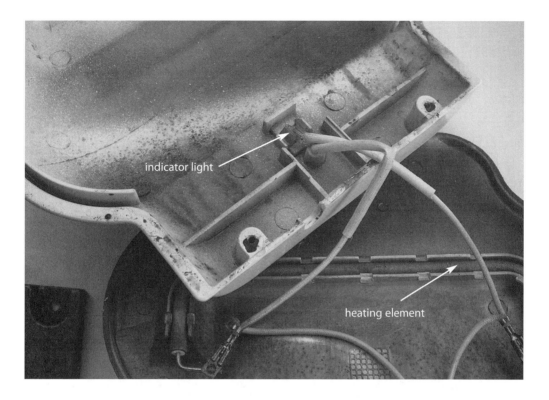

Inside the Electric Grill

A few screws hold the metal grilling plates to the plastic body. With the body removed, the heating wires are visible. The circuit is simple and easy to follow. Power comes into the bottom grill section first. It passes through a thermostat like the one in a coffee pot—you can hear it "pop" on and off. The thermostat is a bimetallic switch: It contains two pieces of metal that expand and move apart when hot, breaking the circuit. When the pieces cool, they come together and close the circuit so the heating element can receive electricity.

The current must also pass through a thermal fuse that protects the user from excessive temperatures. The current contin-

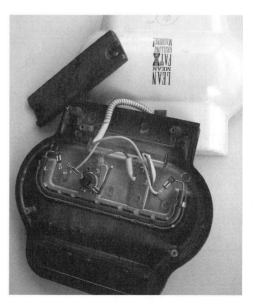

ues on into the lower section's heating element, and then into the upper heating element. A circuit in parallel to the upper heating element powers the small indicator light.

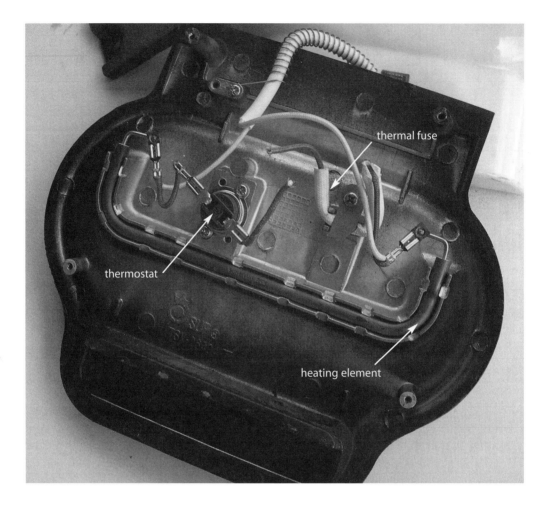

A large block of metal is secured to the base. It helps prevent the grill from tipping backward when the lid is raised.

It Pays the Bills

George Foreman was world heavyweight boxing champion twice. The second time he won the championship, he was 45 years old. As successful as he was in the ring, he has made much more money hawking the George Foreman Grill than he ever did fighting in the ring.

ELECTRIC KNIFE

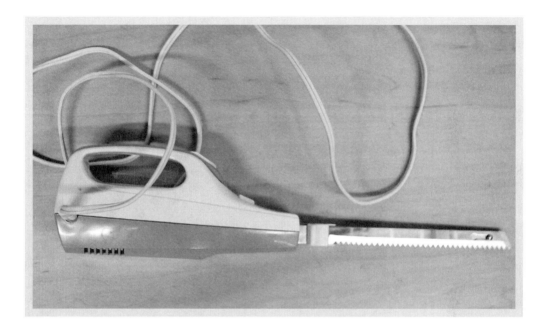

History of the Electric Knife

The trick in making an electric carving knife is figuring out how to convert the rotary (spinning) motion of an electric motor into the reciprocating (back-and-forth) motion needed to cut. Leonard Guilfoyle developed a special gearing system to solve this problem, and patented the reciprocating-action electric knife in 1957 (patent number 2,781,578).

How Electric Knives Work

An electric motor in the handle spins quickly, turning a series of gears. The gears change the direction of rotation 90 degrees and reduce the speed of rotation. Affixed to the gears are cams, components with an off-center rotational pattern that transforms the rotary motion of the motor into reciprocating motion. The cams drive two knife blades mounted side by side and connected to each other at the distant end. A rivet in one blade fits into a slot in the other, which allows the blades to slide past each other as the cams move one blade in and the other out. This back-and-forth motion of the sharp blades is what cuts your turkey.

Originally, electric knives were powered by an electrical cord plugged into a wall outlet. Some still are—including the one dissected in this chapter—but others have rechargeable batteries on board.

Inside the Electric Knife

Unscrewing three screws opens the electric knife. There are two moving parts housed in the top half of the plastic body. One releases the blades for washing or storing and the other is the on/off switch. Both control mechanisms require a downward press to activate. Pressing down on the power switch forces a piece of metal between two contacts so elec-

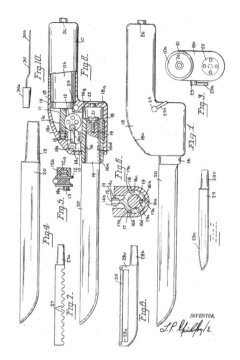

Patent no. 2,781,578

tricity can flow to the motor. When you release the button, a metal spring pushes the switch upward, into the off position.

The two side-by-side blades are held in place by a catch that fits into a slot in the bottom of the plastic drive arm. Depressing the blade release button forces the catch down and out of the slot so you can pull a blade out. Each blade has its own catch. A metal lever returns each catch to its original position when the button is released.

The motor powers both the motion of the blades and a small plastic fan at the other end. The fan draws in air and pushes it past the motor to keep it cool.

A screwlike, spiral-threaded cylinder, known as a worm gear, is affixed to the motor shaft and drives a larger-diameter plastic gear. This combi-

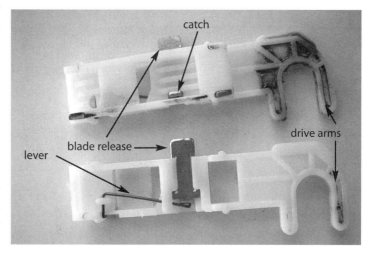

nation of gears achieves two functions. First, it changes the direction of rotation 90 degrees. The motor is aligned with the axis of the knife and it rotates the worm gear along this axis. But the next gear spins perpendicular to the long axis of the knife. A worm gear allows this 90-degree translation of motion.

Second, the gear combination greatly reduces the speed of rotation. Since this knife is powered by alternating current from a wall outlet, its motor spins in sync with the frequency of the electric current: 60 cycles per second, or 3,600 RPM. This is too fast to be useful for cutting and would probably result in too much friction where the two blades are held together. But the gearing system requires the motor shaft to spin eight times in

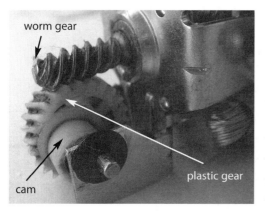

order to get the cams that push the blades to make one complete revolution. So this knife moves the blades in and out at $\frac{1}{8}$ of the speed of its motor—450 RPM, or a bit faster than 7 times a second. If you watch the blades in operation, you can see them move. If they were moving any faster (faster than 10 times a second), their motion would be a blur, as they would be moving faster than your optic system could process the visual signal.

The cams themselves are plastic knobs molded into each side of the larger-diameter gear. They are attached to the gear not at its center but near an outer edge, which gives them their acircular rotation—as the gear spins, they move up and down and side to side along its edge. The plastic drive arms that hold the knife blades fit onto these two knobs. The arms are held in place so they can't rise and fall as the cams move up and down, but they do slide inward and outward as the cams move from side to side. This is how the cams and their connected arms change the motion from rotary to reciprocating.

The cams are located 180 degrees apart on the gear, so that when one moves the left blade outward from the knife handle, the other moves the right blade back.

ELECTRIC TEAPOT

History of the Electric Teapot

An Australian, David Smith, invented the electric kettle and got U.S. patent number 881,968 in 1908. In his design, the heating element was inserted into the middle of the water compartment, in a sealed chamber that kept the water from coming into direct contact with the heating coils. According to his patent application, the coils could be removed and replaced, suggesting that electric heating elements weren't very reliable in the early 1900s.

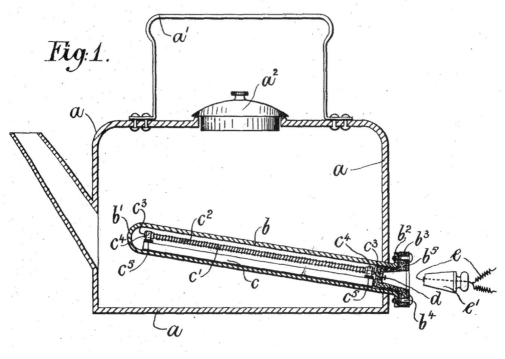

Patent no. 881,968

How Electric Teapots Work

The heating element is made of high resistance wire that transforms electrical energy into heat. (For more information on how heating elements work, see the introduction, p. xiii.) To keep the cost low, there are no switches or transformers. You plug the cord into an electrical outlet and the current flows directly into the heating element. Soon the water is boiling.

Inside the Electric Teapot

Without the tea, there's not much to see. In modern designs, the heating coil usually sits in the bottom of the pot—the electrical wires are inside a metal tube so the water and you don't come in contact with 120 volts. Some newer electric teapots have the heating element in a separate base that the pot sits on.

ESPRESSO MAKER

History of the Espresso Maker

Espresso was invented in Milan, Italy, in 1901 by Luigi Bezzera. Bezzera was trying to invent a faster way to make coffee so his employees would get back to work sooner from coffee breaks. But Bezzera's espresso was brewed at too high a temperature, so it had a bitter taste. Desiderio Pavoni discovered that if he brewed the coffee at a slightly lower temperature—195° F instead of at the boiling point—the coffee tasted better.

Achille Gaggia, an Italian inventor, came up with the piston espresso machine in 1938. The operator pressurized the hot water by pumping it with a piston pump. Instead of Gaggia's hand pump, some machines use electric pumps. Pietro Perucca invented an improved pump for expresso machines, which was assigned U.S. patent number 3,098,424 in 1963.

How Espresso Makers Work

Unlike with regular coffee makers (see p. 20), in which you loosely pile up coffee grounds inside a basket for hot water to fall through, with espresso machines the grounds are tightly packed into a "portafilter," which is then locked in place in the machine. Locking the portafilter is required because hot water will be forced through the grounds at high pressure, which allows the coffee to be ground more finely and packed more densely than with drip coffee.

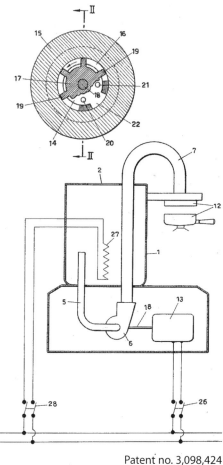

Patent no. 3,098,424

When you turn the switch to "espresso," water is pumped into the heating chamber, where valves close so the pressure inside the chamber will rise as heat is applied. Pressure increases to over 200 pounds per square inch (compared to the pressure of the outside air, which is approximately 14.5 pounds per square inch at sea level). About an ounce and a half of this pressurized water is forced through the coffee grounds in the portafilter and out into your cup. As the brewed espresso burbles from the machine, it has the consistency of syrup. Espresso chemically degrades quickly, so it must be drunk soon after brewing.

To heat cream to add to the espresso, the machine may include a steam wand, under which the user can hold a small cup of cream. With the control knob turned to "steam," an additional heater converts the hot water in the pressurized heating container to steam, which sputters out to heat the cream.

Manual home espresso machines use a spring piston lever, which can deliver hot water or steam at very high pressures. You pull down on the lever to compress a spring; the spring drives a piston that forces the water through the coffee grounds. This design has the advantage of using the spring's consistent force to pressurize the water. If you pushed the piston directly, the pressure would vary with each stroke.

Inside the Espresso Maker

The espresso maker is the tank of the kitchen: heavy and difficult to take apart. At the bottom of this model is a small circuit board that regulates its operation. The user controls are microswitches, small switches that require little force to close or open and that are used

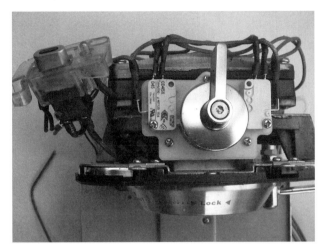

Control switches

in devices as diverse as computer mice and dishwashers. When you turn the dial to select among the espresso maker's several functions, the dial shaft depresses the buttons on two microswitches.

A large, plastic, rectangular bucket on the top of the machine holds water, which feeds by gravity into a pump at the bottom of the machine.

The electric pump in this espresso maker is unlike most other pumps—it works via vibration. The outer cylinder is a large solenoid, a device that creates a magnetic field when a current passes through it. This solenoid's strong magnetic field changes direction 120 times a second, following the change in polarity of the alternating current from the

Pump

wall outlet. Encased in the plastic tube that carries the water is a steel plunger, which the magnetic field draws back and forth as it changes direction. As the plunger moves toward the water line (toward the left in the photo of the disassembled pump), water enters the plastic tube through the hole in the center of the plunger, called a shuttle. When the plunger moves in the opposite direction (toward the right in the photo), the water is forced through the other end of the tube, into the heating chamber.

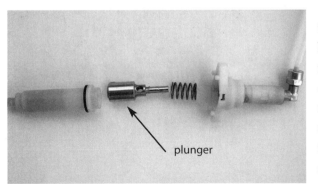

plunger

As the water is forced out of the pump, it pushes a tiny plastic ball away from the mouth of the tube. When the plunger changes direction again, a spring pushes the plastic ball back into position, sealing the mouth of the tube. In this way, it serves as a check valve, allowing water to be pumped out of the tube but preventing it from flowing back in. O-rings form seals between components to maintain the pressure.

This vibrating pump is an ingenuous way to pump up the pressure with very few moving parts. The plunger moves back and forth 60 times a second, pumping a tiny bit of water into the heating chamber each time.

Sitting on top of the pump mechanism and in physical contact with it is a thermal fuse. If the mechanism's temperature rises above 185° F (85° C), the fuse will cut power to the solenoid.

When water is pushed up into the heating chamber, it is pressurized, so the two halves of the chamber are bolted together firmly. The water runs along a

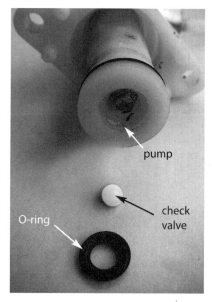

narrow trough, following a circuitous route through the chamber. The heating chamber has three thermal sensors. Two round thermostats sit on the front and back areas of the heater to cycle it on and off. In the middle is a thermal fuse that will cut power if the temperature rises excessively.

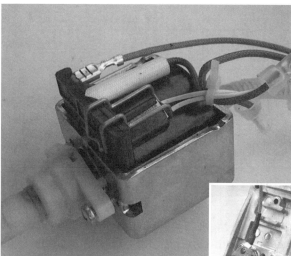

Pump with thermal fuse

From the heating chamber, the hot water can be channeled into the coffee grounds, the steam wand, or an overflow reservoir on the back of the appliance. This reservoir is probably there to allow some cooling before the water is handled; it empties into a tray below.

Inside the heating chamber

Faucet

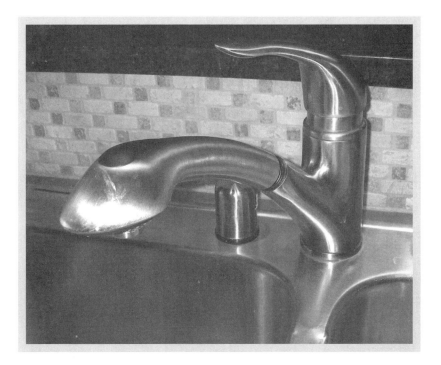

History of the Faucet

As far back as ancient Rome, people have used pipes to move water for cooking. Getting the water to flow where you wanted it was easy, but controlling the flow took much longer to master. Simple taps, like the ones used in outside faucets, use a screw-down mechanism that wasn't invented until 1845. Thomas C. Clarke's tap design was awarded a patent in 1846 (patent number 4,419). Older kitchens and bathrooms use two simple taps, one for the hot water from the water heater and one for the cold water from the water main. Sometimes the two sources are connected to the same spigot, where the water is allowed to mix before gushing out, but each source is still controlled by a separate tap.

Single-handle faucets were invented in 1937. After Al Moen nearly burned his hands when he turned on the faucet, he decided that there had to be a better way. He invented a single-handle design that mixes water from the hot and cold pipes so it comes out at an intermediate temperature. Today there are hundreds of single-handle designs to choose from.

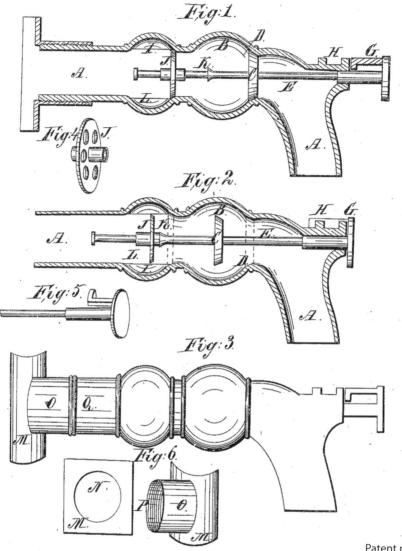

Fig. 1.

Fig. 2.

Fig. 4.

Fig. 5.

Fig. 3.

Fig. 6.

Patent no. 4,419

How Faucets Work

A faucet is the on/off switch and volume control for water in your kitchen. In its simplest form, the faucet handle screws down to seat a rubber or plastic washer over an opening in the water pipe. The screw provides the mechanical advantage needed to stem the tremendous force exerted by water under pressure.

More complex faucets allow one-handed adjustment of the water temperature. A single valve controls the flow of water from both hot and cold water pipes. Lifting the handle pulls up on the valve to admit water to the spout. The farther the handle is lifted, the greater the opening is and the more water comes out. In addition, as you rotate or shift the handle, it moves the valve between the entry points for the hot and cold water pipes, changing the percentage of water coming in from each pipe. In the Al Moen patent drawing shown here (patent number 2,949,933), you can see the inlets for hot and cold water on opposite sides. The more the handle is rotated toward one side, the more water from the pipe on the other side is admitted, thus raising or lowering the temperature of the water that comes out of the faucet. The water enters a "mixer" before exiting the faucet so you get water of a uniform temperature.

On the end of many faucets, you'll find an aerator. This component spreads out the flow of water by passing it through a sieve-like nozzle. Breaking up the water stream releases some of the gases dissolved in the water and mixes the water with air as it falls into the sink.

Inside the Faucet

The illustration below shows the inside of a simple tap. Turning the handle raises the washer-protected stopper and allows water to flow.

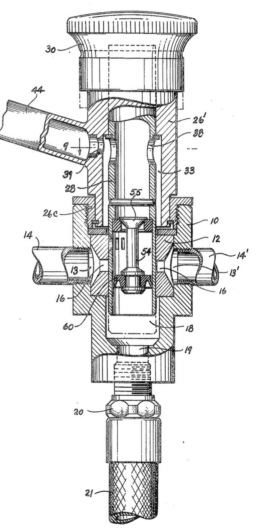

Simple tap

Single-handle faucets, however, hide a more complex interior. The component shown here is a valve for a lever-type faucet. The handle lever screws

Patent no. 2,949,933

Bubbly Water

Have you noticed that sometimes bubbles appear when you get water out of the tap—and that the amount of bubbles changes depending on how fast the water is flowing? At high flow rates, the sudden drop in pressure from the pipes (up to 5 atmospheres of pressure) to the sink (1 atmosphere of pressure) causes the volume of the air bubbles to enlarge by a factor of 5. At lower flow rates, the pressure drop isn't as sudden, so the bubbles may not appear.

into the white plastic section at the center of the valve, while the outer brass structure is mounted into the sink. The cold water pipe feeds one of the two holes at the base of the valve (one hole is visible in this picture), and the hot water pipe feeds the other. Look under your sink to see where the two pipes come together.

Pulling up on the handle raises the center piece while the outer structure remains in place. This action aligns holes in the center piece with the two large holes at the bottom of

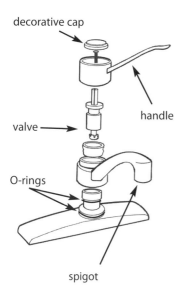

the cartridge, letting water flow in from the pipes. As you rotate the position of the handle, the inner piece rotates and its holes align more with the cold water pipe or hot water pipe.

Single-handle faucet valve

Water from one or both sources flows up through the center of the inner piece, out the smaller holes (only one is clearly visible in this picture, but there are four in all), and into the spigot.

FIRE EXTINGUISHER

History of the Fire Extinguisher

Almon Granger was one of the first Americans to develop a fire extinguisher. He patented a soda-acid extinguisher in 1881 (patent number 258,293). Mixing a water solution of sodium bicarbonate (baking soda, the same stuff in the yellow box in your refrigerator) with sulfuric acid inside the extinguisher generated gas to expel water through a nozzle. Pressing on a plunger broke the container of acid so it could mix with the sodium bicarbonate. Many innovations later, the disposable, dry chemical fire extinguisher was developed, the type found in most kitchens today.

In 1949, Ignatius Nurkiewicz invented the method of using a pressurized gas cartridge to propel the fire retardants out of the extinguisher (patent number 2,533,685). The user pierces the gas cartridge with a pin that is moved by the handle. The gas expands into the main body of the extinguisher, forcing the chemicals it contains out the nozzle.

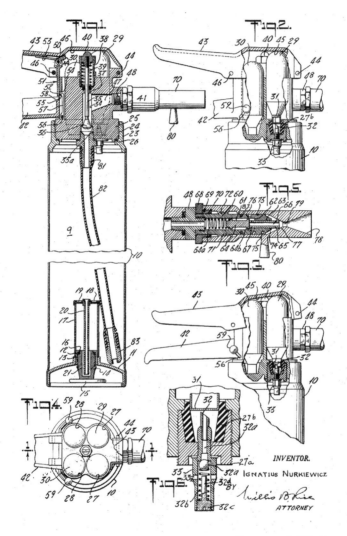

Patent no. 2,533,685

How Fire Extinguishers Work

Most home extinguishers put fires out by covering the burning material and depriving it of oxygen (from the air). Without oxygen, the fire goes out. The material inside the extinguisher is a fine powder of one of three chemicals: sodium bicarbonate, potassium bicarbonate, or monoammonium phosphate. These chemicals are more dense than oxygen and do not react with it—that is, they don't burn. Sprayed on a fire, they prevent oxygen from getting to the burning material.

In addition to the smothering agent, some extinguishers contain a pressurized canister of carbon dioxide, as in Nurkiewicz's design. Squeezing the handle not only pushes down on the rod that punctures the carbon dioxide container but also opens a valve that lets the now-pressurized smothering agent escape through the nozzle. But this design has become less common.

Instead, most modern home extinguishers are "contained pressure" devices. The carbon dioxide propellant is contained not in a separate canister but in the top quarter of the extinguisher itself. Below it, the rest of the volume is taken up by the dry smothering agent. The carbon dioxide gas is pressurized, so opening the valve allows the pressure to push the smothering agent up the siphon tube, out the nozzle, and onto the fire.

Some extinguishers forgo the dry powder altogether, using carbon dioxide itself as the smothering agent. The carbon dioxide gas is heavier than air. Upon release it sinks to the floor and pushes the air up and away from the fire. Deprived of the oxygen in the air, the fire goes out.

Home extinguishers contain only enough material to last 20 to 30 seconds. They are suitable for a grease fire in a frying pan or similar small fires. Larger fires require the response of a fire department.

Inside the Fire Extinguisher

Don't try this at home! Or at the office. To see inside a fire extinguisher, the pressure inside has to be reduced to atmospheric pressure. That is, the extinguisher has to be discharged. And even when it has been discharged, it will still contain some of the fire-extinguishing chemical. It's not dangerous, but you don't want to have to clean up this fine powder.

Starting with a discharged and discarded extinguisher, I unscrewed the nozzle assembly. This piece includes a gauge to tell you if the extinguisher has pressure or needs to be recharged; the valve that regulates the flow of chemicals; the handle and trigger; and the hose and nozzle. Also attached is the plastic siphon tube. This reaches almost to the bottom of the extinguisher.

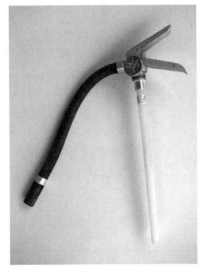

Nozzle assembly

The handle is a lever that allows you to exert some serious pressure to depress the valve. Squeezing the handle pushes down on a pin (valve stem), which opens the valve.

Time for a Checkup?

Periodically check the pressure gage on your fire extinguishers. With time, pressure will be lost, making the extinguisher inoperable. Some can be recharged, but many of the lower-cost extinguishers must be replaced. There's not much worse than needing an extinguisher in an emergency and finding that it won't discharge.

This action is opposed by a metal spring that closes the valve when you loosen your grip. When the valve is open, the dry chemical extinguisher is forced up into the tube, through the hose, and out the nozzle.

Once the nozzle assembly has been removed, you can see inside the tank. Aside from some particles of the dry chemical, it's empty.

Inside view: pressure gauge (left) and valve (center)

Food Processor

History of the Food Processor

Long before food processors were popular in the United States, they were all the rage in Europe. French chef Pierre Verdun invented a series of food processors and obtained a U.S. patent in 1975 (patent number 3,892,365). American inventor Carl Sontheimer improved the European models and created his own version patented in 1976 (patent number 3,985,304). He called it the Cuisinart. Less than a decade later, food processors had largely displaced blenders in American kitchens.

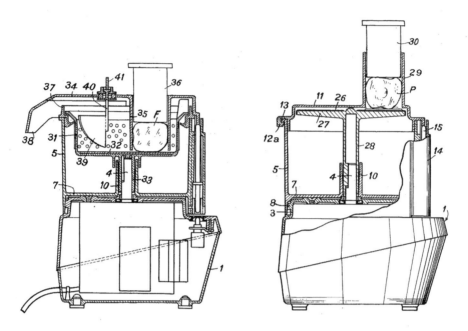

Patent no. 3,892,365

How Food Processors Work

A heavy-duty motor drives the various blades and other attachments at a variety of speeds. A safety lock system ensures that the plastic container is on top and seated in place before the motor can be turned on. You have to turn the mixing bowl, locking it into two plastic hubs, to align the metal rod so it can depress the power button. Inserting the plastic plunger top into the container moves the metal rod down so the motor will turn.

Inside the Food Processor

If you take apart a food processor, you need to be careful of its capacitors, two very large black cylinders that each hold up to 150 volts of charge. Carelessly touching both terminals of a fully charged capacitor will give you quite a shock. To avoid the danger, discharge the capacitor by touching both terminals simultaneously with the metal shaft of a screwdriver—while holding onto only the plastic handle. Touch the terminals of each capacitor several times to make sure you've

touched both at the same time. If a charge remains, you will see or hear a spark. Subsequent touches should not elicit any discharge. Now that you're held in terror by the prospect of being shocked by a vengeful capacitor, keep in mind that an old, discarded food processor will rarely have fully charged capacitors. You should still check, but as long as you make sure they're fully discharged, you will be safe.

Capacitor

The Cuisinart depicted here is a heavy monster. A few screws hold the bottom plastic cover on the base. Removing these reveals the motor and controls. The motor is in the center, connected directly to the spindle that holds the attachments.

The plastic mixing bowl that fits on top of the base has to be aligned properly for the motor to get power. This safety feature is achieved with a spring-mounted lock that depresses a switch on the outside of the base. Underneath is the actual switch, a microswitch. This tiny device allows even a small amount of force to switch power on and

spring-mounted lock

switch

off. It too has a spring-powered assembly. Placing the bowl on the base and twisting it into place are required to align the lock system that allows the motor to receive power.

The user controls, which consist of a button-switch assembly, are held in place with three screws. When you push one of the buttons, it depresses the tiny switch beneath it. Two LEDs let you know if the power is on or the special dough-mixing function is operating.

The master circuit board has three power transistors used in switching power on and off. Because these transistors get hot, they are mounted on heat-dissipating fins. There are also two large black plastic cases holding power relays. These too are switches that control the flow of power. The board also has a transformer that takes your wall outlet's 120-volt electrical power and changes it into the 12 volts the motor requires.

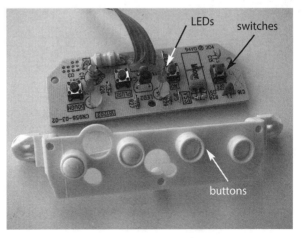

LEDs   switches

buttons

Power for the motor runs through the food processor's two capacitors, a "start capacitor" and a "stop capacitor." Power to each capacitor runs through an electrical choke, a device that removes large fluctuations in the circuit. Many of the electronic devices throughout your house have a choke on their power or data connections. To see one, look through the wires feeding your computer for a fat section of the wire near the end; it will look like a python that has just eaten a rat. The chokes in the food processor are to reduce the large power surges that would otherwise occur when the device, which uses lots of current, is suddenly shut off or turned on.

A start capacitor is used in devices that must be started and stopped quickly—like food processors when you pulse the motor on and off. The start capacitor supplies extra current for the motor when it is first turned on. Since the processor blade might be jammed with dry ingredients, the motor needs some extra oomph to start from zero and get up to speed. Once the motor is nearly up to full speed, the capacitor is removed from the circuit.

Compared to the Cuisinart disassembled above, the Hamilton Beach food processor pictured below is much simpler inside. Whereas the Cuisinart motor is attached directly to the spindle, the Hamilton Beach has the motor off to one side, connected to the mixing system by a series of gears. The small pinion gear on the motor shaft has 12 teeth and the big gear on the spindle has 128, giving a more than 10-fold decrease in rotational speed and corresponding increase in torque, or turning power.

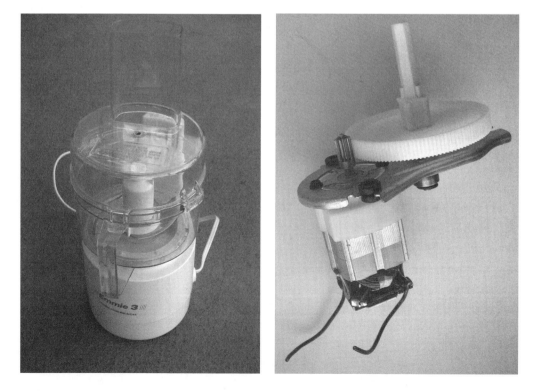

GARBAGE DISPOSAL

History of the Garbage Disposal

John W. Hammes, an architect and inventor, built the first kitchen garbage disposal in 1927. He spent eight years improving the device before making and selling them. He was awarded U.S. patent 2,012,680 in 1935.

From the 1930s to as late as the 1990s, these devices were banned in many U.S. cities. Officials feared that the increased organic load in wastewater would overwhelm the municipal water treatment systems. But studies showed that disposals were not as harmful as was feared, so today most municipalities allow them and about half of U.S. homes have them.

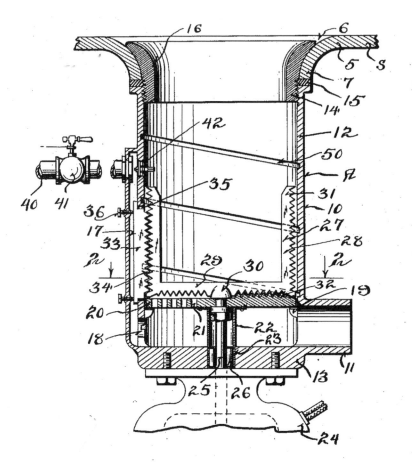

Patent no. 2,012,680

How Garbage Disposals Work

Washing chunks of food down the drain without a disposal is an open invitation to plumb-ing problems. Garbage disposals chew up food scraps into sizes small enough not to lodge in the wastewater pipes. When scraps are dropped into the drain, they land on a metal disk, or flywheel, which holds them as they await disposal. When the appliance is acti-vated—usually by a wall-mounted switch behind the sink—a high-powered motor spins the flywheel on which the food scraps rest, at speeds of about 1,500 to 2,500 RPM.

Attached to the disk are raised pieces of metal called impellers. They don't cut the food—they smash into it. Due to the centrifugal force from the spinning disk, the force of the impellers throws the food outward, toward a stationary ring that surrounds the fly-wheel. This component is called the shredder ring, a version of which is depicted in the illustration from patent number D560,962. The slots in the shredder ring help to chop the food up into manageable pieces.

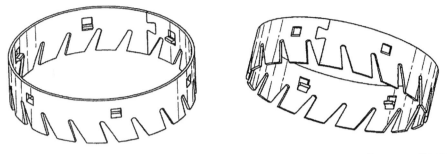

Patent no. D560,962

During operation, water runs through the disposal to pick up bits of chopped food and carry it through holes in the flywheel and the slots in the shredder ring, and down the waste pipes beyond. Some models have an additional cutter, mounted beneath the fly-wheel, that cuts short any long fibers that have washed through the plate.

Inside the Garbage Disposal

Garbage disposals typically have three plumbing connections and one electrical connec-tion. Before messing around with a disposal, ***unplug it***.

Water and food scraps come in from the top, through the sink drain, while the dish-washer pumps its spent water and collected grease and food particles into the side of the disposal. If you don't have a dishwasher, that part of the disposal is sealed shut.

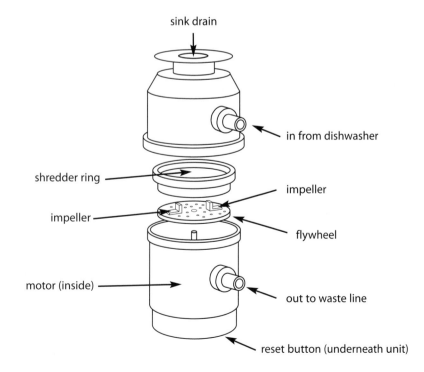

Below the dishwasher connector is the spinning flywheel, surrounded by the shredder ring. Mounted onto the flywheel are the metal impellers that fling stuff into the shredder. They're attached loosely so they can slide around on the top of the fly-wheel, which helps to prevent the spinning mechanism from jamming. Below these components is the pipe that connects to the waste line—this is the output of the garbage disposal. A shaft connects the flywheel to the elec-tric motor, which sits at the bottom of the unit.

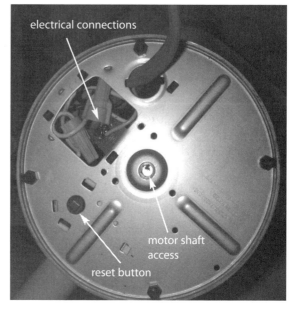

On the underside of the disposal are a reset button, a cover plate for the electrical connections (the plate is removed in the accompanying photo), and access to the end of the motor shaft. If you tossed a T-bone down the drain and jammed your disposal, several of these components may prove use-ful. First **unplug the unit** and try to remove the bone from the top—preferably with tongs, if possible. If that fails, find a wrench that fits the end of the motor shaft and try turning it by hand to dislodge the bone. Once the offending item has been removed, plug the dis-posal back in and push the reset button. That whirring sound means you don't have to call a plumber or replace the disposal.

Why Cold Water?

When grinding garbage, always flush the disposal with cold water. Water washes away food particles, preventing them from building up and clogging the drain. Cold water helps solidify fats so they can be broken up mechanically by the disposal. Warm water might melt the fat in the disposal, allowing larger pieces to pass through into the waste pipes below, where they can resolidify and clog your plumbing.

Garlic Chopper

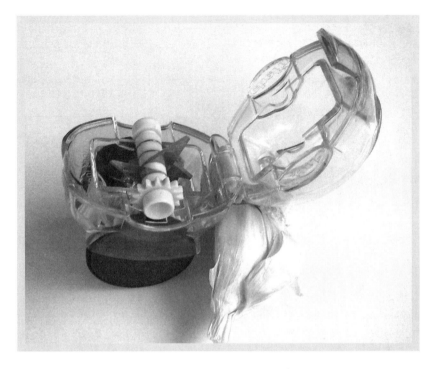

History of the Garlic Chopper

Adam Jossem and David Holcomb patented the Garlic Zoom garlic chopper depicted in this photo in 2008. It is listed as a "food processor" in the design patent (patent number D576,836).

The garlic press was patented earlier, in 1954, by Steven Sarossy (patent number 2,776,616). A more traditional design, it presses the garlic and juice through small holes. Sarossy's was the first garlic press recognized by the U.S. patent system. (Older designs existed but weren't patented.) G. A. Lacout invented another type of vegetable chopper in 1950 (patent number 2,507,571), intended for mincing garlic and other herbs.

How Garlic Choppers Work

With the Garlic Zoom model of garlic chopper, you drop a clove in the top, close the lid, and run the wheels along your kitchen counter. The wheels turn four blades, which mince and dice.

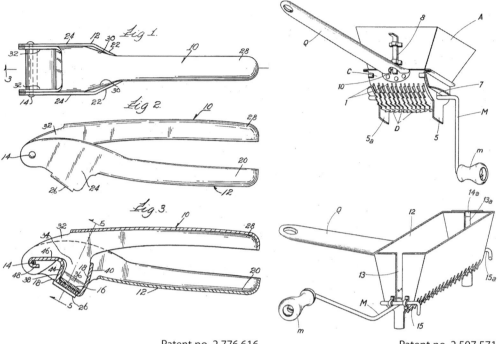

Patent no. 2,776,616   Patent no. 2,507,571

Using a garlic press instead of a chopper is easy, but cleaning the press is more difficult. Many later inventions have focused on the problem of how to get all the bits of garlic out of the holes in the press. Some presses come with a plastic comb with bristles that align with the holes in the press to force out lingering garlic chunks.

Inside the Garlic Chopper

Nothing much to disassemble here—simply open the lid to access the cutting blades. They're connected to a shaft that lifts right out of the housing.

Beneath them, the two wheels ride on an axle that has a large gear. This gear drives a smaller gear on the cutting shaft. The gear combination speeds up the rate of turning by more than two to one. Two inches of countertop revving turn the blades one full revolution inside. The blades can slide slightly on the shaft, presumably so they don't get damaged if they encounter something too hard to cut.

And that's all there is to it. Pretty elegant design for a gizmo that's fun to use and simple to clean.

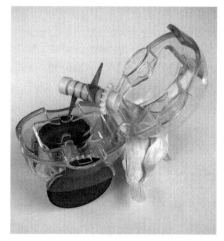

GROUND FAULT CIRCUIT INTERRUPTER (GFCI) SWITCH

History of the GFCI Switch

Charles F. Dalziel invented the GFCI, which he called the "miniature differential circuit breaker," in 1965 (patent number 3,213,321). Dalziel was a professor of electrical engineering at the University of California, Berkeley, whose research was on the effects of electricity on people and animals. His invention was designed to prevent electrical shock by shutting off power if a short circuit occurred. When he was unable to interest investors in the device, he reputedly demonstrated its utility by tossing an electric toaster into a bathtub where his eight-year-old daughter was sitting. It can be a tough life being the child of an inventor.

The specific GFCI outlet disassembled in this chapter was patented in 2005 by Nicholas DiSalvo and William Ziegler (patent number 6,864,766).

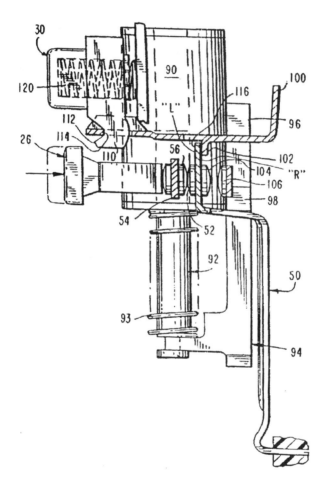

Patent no. 6,864,766

How GFCI Switches Work

In the kitchen, GFCI switches are most often found as part of electrical outlets adjacent to sinks. A GFCI switch shuts off the flow of electricity in cases in which the current may have left its proper path and started flowing through *you* instead. This can happen when a misused or faulty appliance puts you in contact with an electrically charged wire and you complete a circuit between that wire and a ground. For example, if you are turning on the sink and you reach for a toaster that's malfunctioning, electricity could flow from the toaster through you to the cold water pipe. The pipe acts as a ground and you could be shocked. (Appliance manufacturers fear this scenario and design their wares so it won't happen. But it can, and infrequently it does). But as the electricity rushes to the ground, the GFCI senses the sudden change in current and shuts off the circuit, protecting you from potentially serious injury or even death.

Inside the GFCI is a device called a differential current transformer, which compares the current flowing into the plugged-in device with the current flowing back out of it. If the flows are different, current must be leaking from the device into a ground, possibly through the person using the device. If there is an imbalance (of as little as 5 milliamps), the GFCI interrupts the current within a few milliseconds.

The Electric Safety Foundation International says that GFCIs have "saved hundreds of lives and prevented thousands of injuries in the United States over the last 30 years."

Inside the GFCI Switch

With the plastic housing removed, the electrical connections are visible—note the screw connectors at the upper and lower right. Incoming current flows in from the electrical supply at the top, and outgoing current flows back out to the electrical outlet at the bottom. The long copper fingers in the center are contact arms that carry the electricity. A ground-wire connection is on the left side.

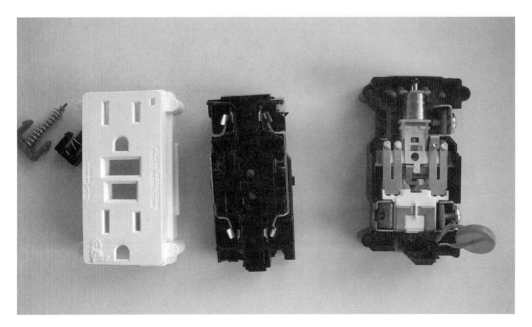

Some industry experts estimate that as many as 200 electrocutions a year could be prevented if everyone used GFCI switches when operating electric appliances near cold water pipes.

The shiny copper coils at the top make up a solenoid, a device that creates a magnetic field. When the differential current transformer detects an imbalance in the power flow, the magnetic field in the solenoid increases, which pushes a plunger to cut off the flow of electricity from the electrical supply to the contact arms. To the right of the solenoid is an indicator LED.

Behind the components pictured here is a circuit board, to which the large disk at the lower right is connected. This disk is the differential current transformer. Inside are two coils, one connected to the incoming line and the other connected to the outgoing line. Each one generates a magnetic field based on the power flow in that line. An integrated circuit on the circuit board detects any imbalance between the magnetic fields of the two coils and triggers the solenoid.

The solenoid can also be triggered by depressing the "test" button, which pushes down on the silver toggle and trips the circuit. The reset button, on the other hand, depresses a switch in the center through the round opening, which restores power after the circuit has been tripped.

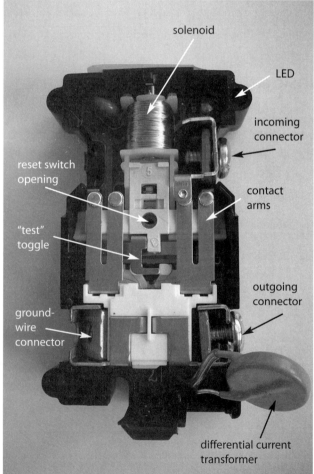

solenoid

LED

incoming connector

reset switch opening

contact arms

"test" toggle

outgoing connector

ground-wire connector

differential current transformer

HOT PLATE

History of the Hot Plate

It's not clear who first thought of inventing an electric hot plate. Designs for electric cooking machines received patents starting in the late 19th century. Many improvements were made, including this one that was assigned to General Electric in 1941 (patent number 2,255,500).

How Hot Plates Work

The heating element is made of Nichrome high resistance wire, which converts electricity into heat. (For more information on how heating elements work, see the introduction, p. xiii.) In the hot plate shown, the heating element has about 12.5 ohms of resistance, which means that its maximum current is 9.6 amps. (Divide the 120 volts provided by the

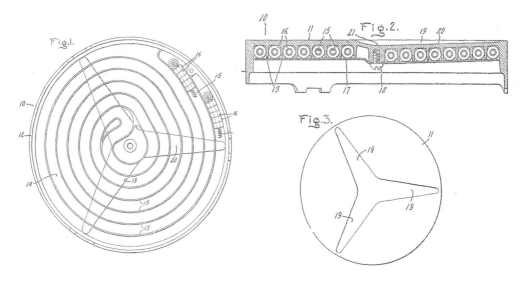

Patent no. 2,255,500

wall outlet by 12.5 ohms of resistance to get the current.) That current multiplied by the voltage (again, 120 volts) gives the power rating for the hot plate—1,152 watts, which is about the same as the official marked rating (1,100 watts).

Inside the Hot Plate

I took apart a Sears Table Range to see what lies inside. Electrical power is provided to the heating coil through a thermostat. You turn the control dial to set the temperature of the burner. However, the thermostat isn't calibrated; the numbers on the dial aren't temperatures, just levels of heating.

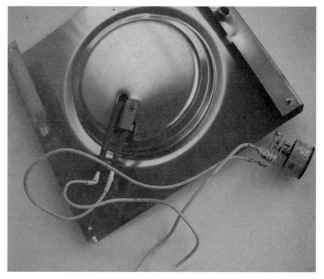

As with many devices that have temperature controls, this hot plate's thermostat consists of a bimetallic strip that changes shape when heated. The strip holds the first of two electrical contacts. When the temperature dial is turned to the off position, a bent metal spring pushes a second contact point against the first, which cuts off the flow of current to the heating element. As you turn the control dial clockwise, a screw moves the second contact away from the first, providing power to the burner. But as the bimetallic strip heats up, it bends, moving the first contact back toward the second and cutting off the heat. Turning the dial to its highest possible setting gives the maximum possible separation between the two contacts. In this position, the bimetallic strip will have to bend far to bring the two contacts together—which means higher temperatures are required before the burner shuts off.

ICE MAKER

History of the Ice Maker

The technology for making things cold has always lagged behind the technology for making things hot—it's just easier to heat something than to cool it. The first automatic ice cube maker for refrigerators was invented by Robert Galin in 1958 (patent number 2,846,854). In Galin's design, cubes were freed from the ice cube mold by "tilting and distorting" it with an electric motor. Later improvements used heat to melt the ice along the mold so it could be pushed out by mechanical fingers.

How Ice Makers Work

Where does the water come from? The ice maker is connected to a cold water pipe, sometimes by fitting the pipe with a clamp-and-screw assembly called a saddle valve. If you go

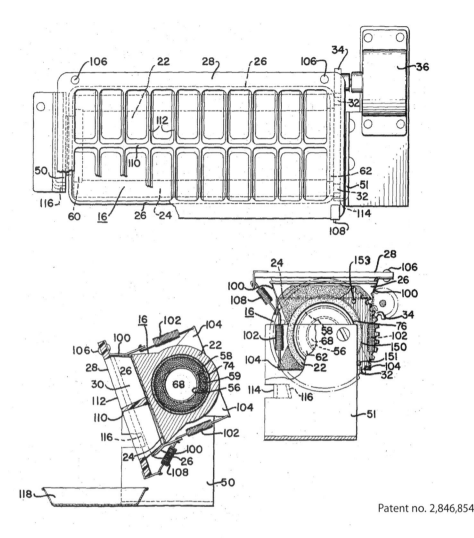

Patent no. 2,846,854

looking through your house you may find several of these copper devices, one supplying water for the refrigerator and another feeding a humidifier. A saddle valve can be installed anywhere on a pipe with two sets of nuts and bolts. With a few twists of its "T" handle, it advances a sharp point, puncturing the pipe and allowing water to flow into the copper tubing of the valve itself. From there it goes into the plastic tubing that connects to the ice maker.

Water flows through a valve at the back of the refrigerator into the ice cube mold. The valve opens for only a few seconds (about seven) to fill the mold and then shuts off. You can hear the valve open and close. The water freezes into ice cubes in the mold. When a thermostat senses that the temperature in the mold is sufficiently low, it signals a heating element to come on briefly. The heating element warms up the bottom of the ice mold so the ice can be removed. Then a motor spins a series of gears connected to a row of plastic fingers, one in each ice cube slot of the mold, to push the cubes out of the mold and into the bin below. (There, of course, they freeze to each other, making it next to impossible to extract individual cubes.)

Once the fingers have pushed the cubes out, they trigger the system to start over again by filling the mold with water. If the shutoff arm is raised, the shaft the fingers are mounted on cannot turn, so the ice stays in the mold and the restart function is halted.

Inside the Ice Maker

Removing the ice maker requires you to unleash the machine from its wiring harness. A clip holds the harness into a bracket on the ice maker; depress the clip to set it free. The front cover pulls off to reveal the control module, with its plastic geared wheel.

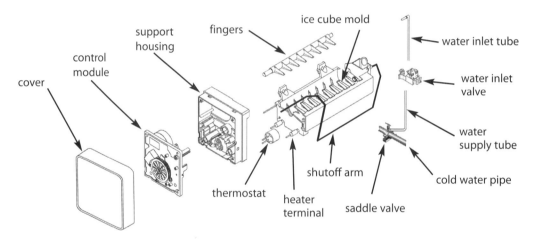

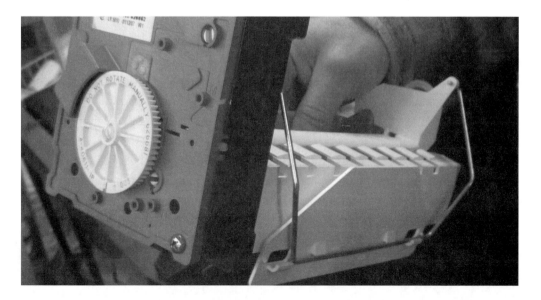

Three screws hold the module in place, and by removing them you can look at its reverse side. There you'll find the ice maker's motor. It is mounted on top of a set of gears, and the entire assembly is screwed to the back of the module. The final gear protrudes through the module to drive the large plastic gear on the front. As the large gear spins, it turns the opening for the drive shaft, visible on the reverse side. The drive shaft, in turn, rotates the fingers to pop out the ice cubes.

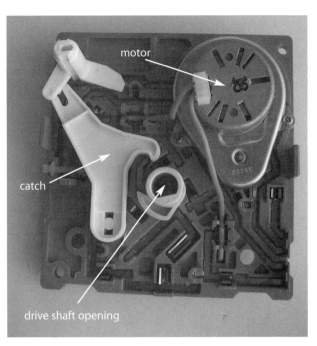

Also visible on the back of the module is the mechanism that prevents the fingers from rotating. When the ice maker's shutoff arm is raised, it rotates the catch on the back of the module. The catch stops the shaft from turning; ice is made in the tray, but not dumped.

Behind the control module is the plastic support housing. Several metal rods protrude from the housing, providing connections between the control module and the rest of the ice maker assembly. The two rods at the bottom center are contacts for the thermostat. The two at each side are the contacts for the heating element. Directly above the thermostat is the end of the drive shaft that the fingers are mounted on.

Two more screws hold the housing to the rest of the assembly. Removing them gives views of the remaining components. The heater contacts are connected directly to the heating element on the bottom of the ice cube mold, but the thermostat is a separate device embedded in the housing. Pushing on the thermostat contacts pops it out.

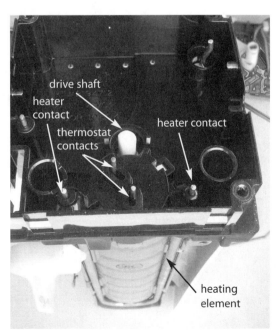

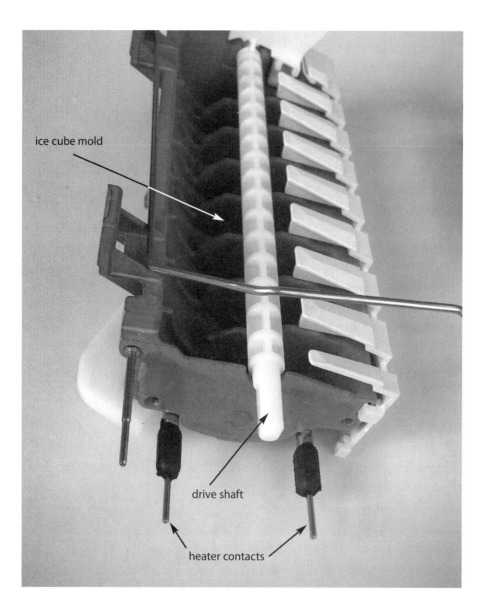

ice cube mold

drive shaft

heater contacts

Freezer Sounds

The process of making ice cubes is acoustically rich—you can hear much of its operation. You hear the valve opening to admit water and closing a few seconds later to shut the flow. The water pipes might also reverberate when the valve closes. The motor turning the cubes out of the mold has a weak grr-ing sound, and then the cubes hit the collecting bin with a crash.

Juicer

History of the Juicer

Hand operated juice squeezers have been used for centuries. The first person to patent a new model in the United States was L. S. Chinchester, in 1860 (patent number 28,967). His design for a "lemon squeezer" is similar to those found in kitchens today. It uses a lever arm to apply pressure to the lemon.

In 1910, Leslie S. Hackney was awarded the first U.S. patent for an electric juicer (patent number 968,344). His design allowed people to operate the juicer with either hand power or an electric motor. He foresaw its use in "drug stores, saloons, fruit stands and places on the street where orange juice and lemonade are dispensed."

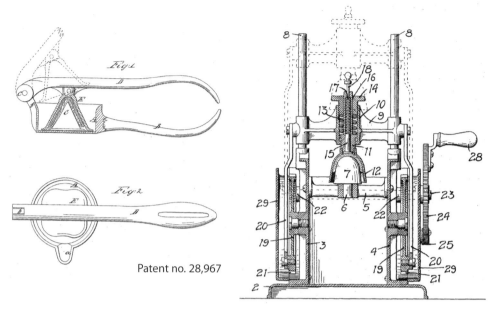

Patent no. 28,967

Patent no. 968,344

In 1950, the Hewlett Manufacturing Company was awarded a patent for an electric juicer designed for home use (patent number 2,515,772). The machine, invented by Raymond Hewlett, automatically cut and squeezed oranges and other fruit.

How Juicers Work

Juicers apply pressure to or mash the pulpy parts of fruit to free the juice and then strain the juice from the other parts. Manually operated machines often use levers or gears to multiply the force the user is able to apply.

A simple juicer is found throughout Southern Asia, where street vendors sell sugar cane juice. The operator places a cane stalk between two geared rollers and uses a large lever to press the rollers together around the cane. He or she then pulls the cane through the rollers, which squeezes out the juice. Juice falls into a collecting top and funnels into a reservoir and then into a glass.

Some citrus juicers accept fruit that has been cut in half, compressing each half to squeeze out the juice. These juicers can be manual, operated with a long lever, or electric, with a motor that provides the power to squeeze.

More common in American kitchens are electric juicers. As you insert fruit into the intake chute, a motor spins a cutting disk that chews apart the fruit. The disk is attached to a basket filled with holes to strain the juice. The juice flows down through the holes, out the back of the juicer, and into your waiting glass. The pulp, on the other hand, overflows the basket as you jam more fruit down the chute and is forced out the front of the juicer.

Choice of materials is especially important in making juicers. The high acidity of many fruits can quickly corrode some components, so higher-quality juicers use corrosion-resistant materials.

Inside the Juicer

The top comes off by releasing two hand latches. This allows you to remove the cutting basket and clean the inside. This model's cutting disk features nine lines of cutters.

Four screws hold the bottom of the chassis to the motor assembly, and three more attach the motor to the top of the chassis. Removing these screws gains access to the motor and controls. The plastic turntable that rotates the cutting basket above is screwed onto the motor shaft. Unscrewing it lets the motor fall out and sets the turntable free. The motor drives not only the cutting basket on top but also a small fan below. The fan circulates air to prevent the motor from overheating.

Cutting basket

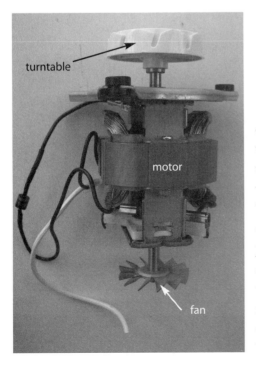

turntable

motor

fan

The on/off switch is a plastic slide on the outside of the chassis that connects to a plastic arm on the inside. As you push down on the slide and depress the arm, it makes contact with the button on a microswitch beneath it. This tiny switch makes and breaks the electrical circuit to operate the juicer. If your juicer just quits running, this is the easiest and cheapest possible repair to try. Simply pull out the microswitch and check to see if it is working—use an electricity meter to see if it passes electricity and then stops the flow when the button is toggled on and off.

On one of the electrical power leads to the motor is a silicon-controlled rectifier, or diode. This device converts the alternating current supplied by the wall outlet into the variable DC voltages required by the motor.

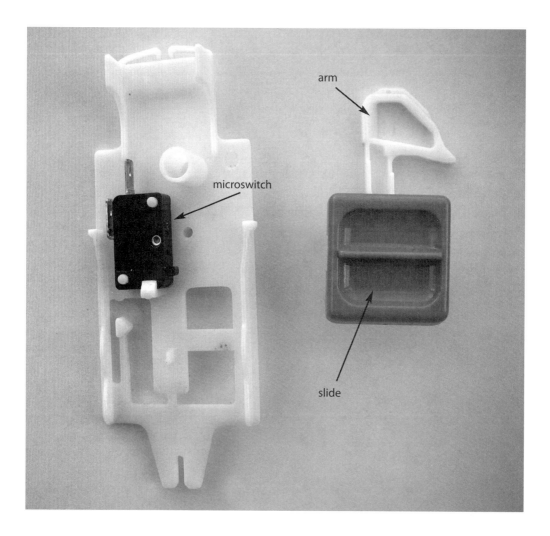

KITCHEN TORCH

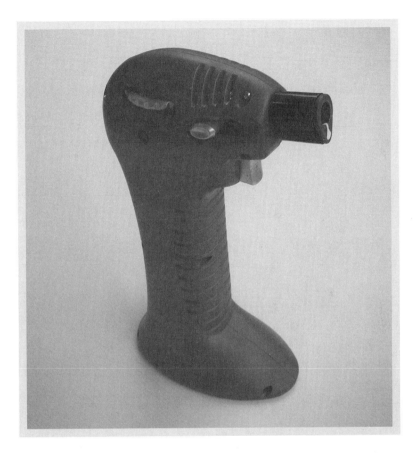

History of the Kitchen Torch

Culinary torches were an obvious outgrowth of the butane and propane torches used in industrial settings to solder pipes, strip paint, etc. Because the mechanisms in industrial and kitchen torches are the same, a standard utility patent was probably not issued to recognize the initial development of this new use. Instead, design patents for the external or aesthetic design of individual models and utility patents for improvements to the device's function would be issued. For instance, Chi-Sheng Ho from Taiwan was awarded a patent in 2001 for an "active safety switch for gas burner" that makes the kitchen use of a hand-held burner much safer (patent number 6,196,833).

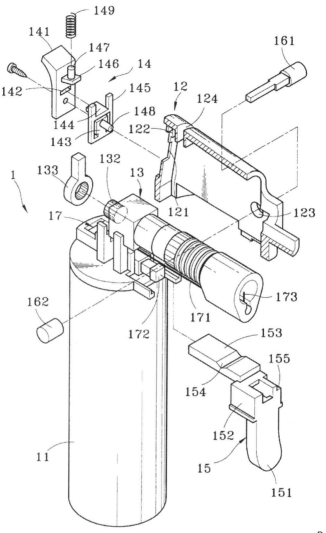

149
141
147
146   14
142   145   12
       148   124   161
144   122
143
132
133   13
1
17   121   123
162
172   171   173
11   153
155
154
152
15   151

Patent no. 6,196,833

How Kitchen Torches Work

A pressurized canister of butane fuels the torch. Depressing a trigger or turning a dial opens a valve, letting the gas squirt out of the nozzle. The stream of gas pulls air in through holes in the tube leading to the nozzle. Adjusting the air flow through the holes is a secondary way of adjusting the intensity of the flame.

The ignition system takes advantage of the piezoelectric effect, by which certain materials—semiconductor crystals—generate electric sparks when subjected to mechanical stress. The spark that ignites the butane arises when a hammer strikes a crystal inside the igniter.

Butane-fueled torches burn at a temperature of about 2,500° F. Fired at a sugarcoated dessert, the flame caramelizes (or oxidizes) the sugar. (Caramelization of sugars occurs at between 230 and 320° F.) In the caramelization process, hundreds of chemical reactions occur, including the transformation of sucrose into fructose and glucose.

Inside the Kitchen Torch

Disassembling a kitchen torch is not dangerous—provided you do not try to open the gas canister itself. That could be bad.

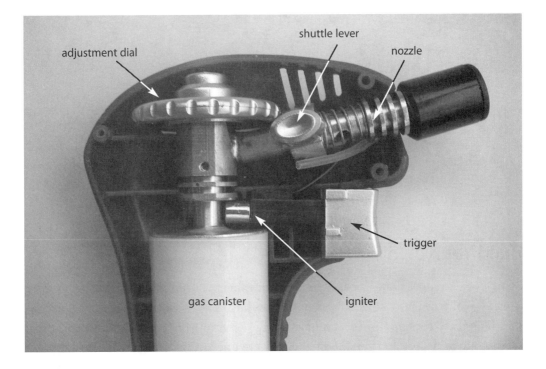

A handful of tiny screws hold the two halves of the body together. Lifting off one half reveals the gas canister, which takes up most of the interior space. On top of the canister is a valve that is opened and adjusted with an "adjustment dial." With the valve open, butane flows out through the metal pipe that leads to the nozzle.

The nozzle is a metal tube with holes cut in it, wrapped in a spring. The operator can push the "shuttle lever" toward the nozzle to close the holes in the tube. This adjusts the size of the flame by controlling how much air is drawn into the pipe along with the butane gas. When the shuttle lever isn't being pressed, the spring returns it to its original position to keep the holes open.

The trigger and igniter sit below the pipe and nozzle. Pushing in the trigger compresses the return spring. It also compresses a spring inside the piezoelectric igniter cylinder. When the interior spring has been compressed sufficiently, a catch releases it so it can strike against the quartz crystal that is also contained within the igniter cylinder. Striking the crystal generates a spark that is carried by wire to the nozzle.

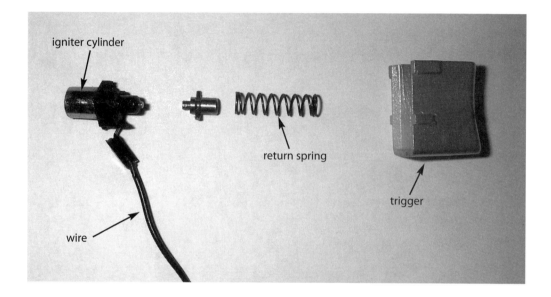

Build Your Own

Some people report that propane torches from their plumbing kits make great crème brûlée—with no taste of solder. Propane torches burn slightly hotter than butane torches, meaning faster cooking times, and have a huge reservoir of fuel. Not to mention that they're much cheaper and you might already have one lurking in your basement.

KNIFE

History of the Knife

Before people were even human (*Homo sapiens sapiens*), they were using knives. Starting about 2 million years ago, they sharpened flint or animal bones or create rudimentary knives, without handles. But it wasn't until about 5,000 years ago, with the discovery of metals, that knives as we know them were invented. The various metals, first bronze (3300 B.C.E.) and later iron (1500 B.C.E.), were used until better materials were discovered.

People began making knives out of steel starting in around 1400 B.C.E., and the practice continues to this day. Nevertheless, the quest continues to find materials that will provide a stronger, sharper, longer-lasting blade.

J. W. Androvatt and T. W. Joline were awarded one of the first U.S. patents for the design of a kitchen knife (patent number 114,388). Their improvement to this age-old device was the curved end that can be used for "scraping out the corners of pans or other vessels."

How Knives Work

It sounds so simple, but do you know how knives cut? There are two different forces at work. First, a knife transfers and focuses your downward force into the small, sharp edge of its blade, which is able to break through that tough carrot. The blade squeezes and separates the carrot while the rest of the knife acts as a wedge. The sharper the knife is, the smaller the area into which your force is focused, and the easier it is to push and cut through.

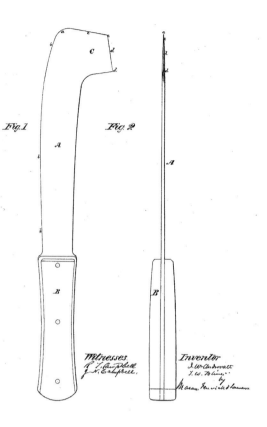

Patent no. 114,388

Second, the edge of a knife is covered with many tiny serrations or burrs. As you pull and push the knife, these serrations tear the food, like a wood saw tears the fiber of a pine plank. A sharp blade has more tiny serrations than a dull blade—another reason to keep those knives sharp.

Some knives also have larger serrations. For instance, serrated vegetable and bread knives have notched blades designed to break through the outer, tougher surface without smushing the softer interior. Knives for other cutting tasks have different cross-sectional shapes, or "grinds." A hollow grind, where the blade has a convex shape on both sides that tapers down to a very thin, sharp edge, is used for the general kitchen knife.

Before forks were invented in the 10th century, knives and spoons were the principle utensils for eating.

There are dozens of different knives just to cut meat. A cleaver is heavy so its downward momentum carries it through a thick cut of meat. A carving knife is thin so the user can cut thin slices. Filet knives are flexible, while most boning knives are stiff. Knives for opening clams and oysters are short and have thicker blades to pry apart the bivalve shells.

Have you seen knives with dimples along the edge? This type of knife, invented by the Granton Knife Company in 1928, has oval divots along its edges. The divots alternate on each side. This design is currently gaining popularity as the *santoku* knife. *Santoku* is a Japanese word that means "three virtues," and this knife has three purposes: chopping vegetables, slicing fish, and cutting meat. The scalloped edge provides air space between the blade and food being cut. This, the chefs tell us, makes cutting easier and decreases the need for sharpening.

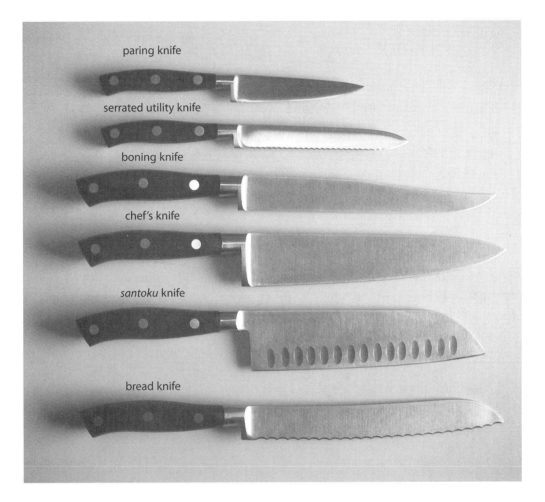

Inside the Knife

Knives have parts just like other kitchen appliances do. One end of the steel form is encased in a wooden or plastic handle called the *tang*. To provide more weight to the knife for better balance, knives have a collar between the tang and the blade called the *bolster*. Some bolsters have finger guards to keep fingers from slipping under the cutting edge.

As for the blade itself, the end nearest the tang is called the *heel*. The

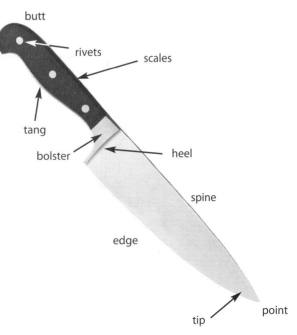

Taking Care of Knives

Knives need tender care in washing. They should be washed and dried by hand, not left to drip dry in a dish rack. Wash them after dinner rather than letting them soak overnight. Long-term baths can make knives susceptible to rusting. Don't use steel wool, because it scratches the blade. When clean and dry, store knives carefully either in a knife block (a wooden stand with slots for each knife) or on a magnetic strip.

opposite end of the blade is the *tip* and the very end of the tip is the *point*. The tip is used for delicate or precise cutting, while the point is for piercing. The full cutting surface of the blade is the *edge*. Opposite to the edge is the *spine*, which gives strength to the knife without increasing the width of its cutting edge.

MICROWAVE OVEN

History of the Microwave Oven

Engineer Percy Spencer discovered the cooking effect of microwave radiation by accident. The story is that while experimenting with microwaves, he discovered that a candy bar in his pocket had melted. When he associated the melting with the microwaves, he tried popping popcorn. When that worked, he made the world's first microwave mistake: trying to cook a raw egg. It exploded, and Percy's mind exploded with the possibilities of microwave cooking. He was awarded a patent in 1950 (patent number 2,495,429).

Raytheon brought the first microwave oven to market in 1947, but it was a monster that weighed 750 pounds, cost a fortune, and used three times as much electrical power as models do today. Not until 1967 was a popular model introduced.

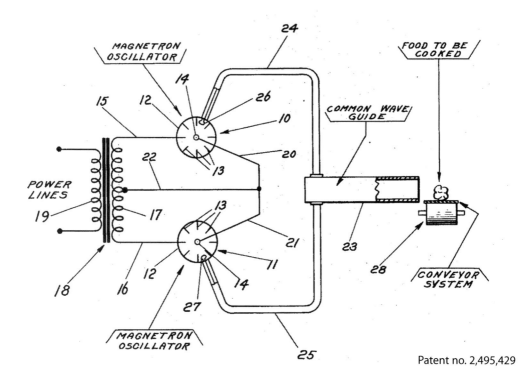

Patent no. 2,495,429

How Microwave Ovens Work

The microwave bombards certain molecules inside food with electromagnetic radiation to heat it. The radiation has a wavelength of about 12 cm (5 in.) and can pass through glass and transparent plastic much like visible light can. Because of this, the glass cover on a food dish doesn't interfere with cooking the food. Similarly, the glass viewing area in the microwave door provides no protection from the radiation—which is why the metal screen is attached to the inside of the glass. Its openings are much smaller than the wavelength of the electromagnetic radiation, so it blocks the escape of microwaves but allows visible light (which has a shorter wavelength) through.

Modern microwave ovens are energy-efficient machines. In traditional ovens, 50 percent or less of the heat generated actually heats the food, but microwave ovens operate at 70 to 80 percent efficiency.

Water, fats, and sugar molecules, however, absorb the radiation, which causes them to vibrate back and forth. They vibrate because their molecules are not symmetric—for example, the two hydrogen atoms in H_2O are not 180 degrees apart but only 105 degrees apart. An asymmetric structure gives each molecule a positive end and a negative end, which will align with the electric field set up by the microwaves. The electric field created by the microwaves changes millions of times each second, so the molecules are constantly realigning themselves. Their back-and-forth motion generates the heat that cooks the food.

To generate the microwave radiation, the oven uses electricity to power a device called a magnetron, which is a kind of vacuum tube. Vacuum tubes are glass-encased devices that emit a controlled flow of electrons. Before transistors and solid-state components, electronic circuits were dominated by vacuum tubes. Television sets had dozens of them, and supermarkets had tube testers so you could find the bad tube in your TV and replace it. Were those the good old days?

The magnetron requires high-voltage direct current (2,000 to 3,000 volts). To get the voltage that high, the oven includes a massive transformer, which constitutes much of the appliance's weight. Inside the magnetron, voltage is applied to a filament, which generates a stream of electrons down the inside of the metal tube. As the electrons zip from the filament, circular magnets force them to spiral. The spiraling beam speeds by cavities in the side of the tube, inducing a high-frequency radiation that is emitted from the end of the tube. Your leftovers are now ready.

In some models a stirrer rotates above the cooking food, turning about once a second. Its fanlike blades interrupt the flow of radiation and help scatter it throughout the cooking chamber. You can tell if your microwave has a stirrer by watching for it (or its shadow) when you open the oven door. The fan blades will continue to spin (and interrupt the interior light) for a few seconds.

Inside the Microwave Oven

Don't do this at home. There are some cool things inside the microwave—and one that can hurt you badly. If you insist on disassembling a microwave oven as I did, you must be

careful of its huge capacitor, which stores electrical energy even when the device is unplugged. Discharge it by touching both contacts with a metal screwdriver, while holding it only by its plastic handle. Touching both contacts with your finger would subject you to a dangerously strong electric shock.

On the oven pictured here, a few screws held in place the metal shield that protects the capacitor. I removed the

shield, found the capacitor, and discharged it. There were no sparks, suggesting that this oven had not been used in a long time and that any stored charge had already dissipated. The capacitor is enormous! It is bright and silver colored, about the size of a pack of cigarettes. It stores electrical energy so it can build up to power the magnetron. The capacitor is connected to the chassis through a rectangular diode, presumably to channel any electrical charge from the chassis to the capacitor, in addition to heavy power cords that connect it to the transformer and the magnetron.

The magnetron is mounted in the chassis near a thermal fuse. If something goes wrong and the oven seriously overheats, this fuse will cut power to the generator. A second thermal fuse monitors the temperature in the cooking compartment.

The magnetron looks like the end of a microscope. The tube is the waveguide that directs the radiation downward, into the cooking chamber. The waveguide is surrounded by cooling fins and the two powerful magnets that guide the electrons emitted by the magnetron. So powerful are the magnets that when extracted from the oven and allowed to join each other, they are a struggle to separate. Keep them away from your watch, as they could bend the delicate metal parts and make time (for you) stand still.

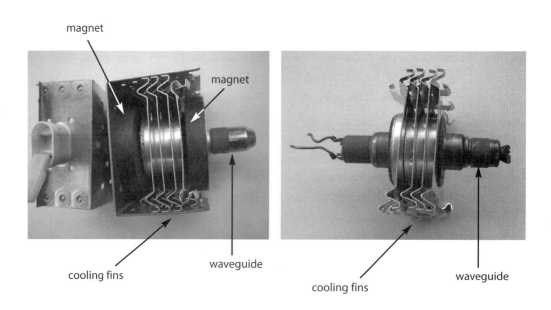

magnet

magnet

cooling fins

waveguide

cooling fins

waveguide

At the bottom of the cooking chamber sits a carousel, which spins the food being cooked to ensure that the microwaves are evenly distributed throughout it. The carousel is turned by a motor mounted to its underside. It runs on 120 volts of AC power. The motor is attached to several gears—it takes a pair of pliers and significant force to turn the final gear.

This microwave oven doesn't have a stirrer to spread the microwaves about, but it does include a fan that both cools the magnetron and mixes air inside the cooking compartment. A metal fin divides the flow of air between these two functions.

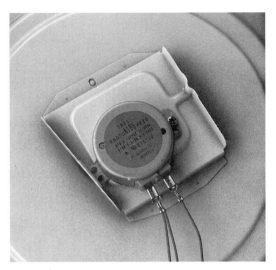

Carousel motor

The door latch assembly is fitted with three microswitches to ensure that the door is shut when the microwaves are firing. These tiny switches appear in many devices, including computer mice. In many large appliances they are used as safety switches; they prevent the power from coming on unless all the components are in the correct position. In this microwave, the three microswitches are closed by the two plastic fingers that extend from the door and fit into the latch.

microswitches

The control panel is one long circuit board. On the side facing the user are touch buttons to control the operation, and at the far left is the LCD readout for the timer. On the reverse side is a large piezo speaker; it contains a crystal that produces sound in response to changing electrical voltage. That's what makes those annoying beeping sounds when the timer runs down. At the other end of this side are a transformer, which changes the voltage from 120 volts provided by the electrical outlet to the much higher voltage required by the magnetron, and a relay switch, which controls the flow of power to the transformer.

relay switch

piezo speaker

transformer

Mixer

History of the Mixer

What we call mixers today were once called eggbeaters. In 1870, Turner Williams patented the first one that kitchen chefs would recognize today (patent number 103,811). The design has been very successful, and similar products can be found in kitchens today.

Michael Hall added an electric motor to an "egg-whisker and cake-mixer" in 1909 (patent number 917,289).

How Mixers Work

With a handheld mixer, you turn an upright wheel with your wrist and a set of gears rotates that motion 90 degrees, translating it into the spinning motion of the beaters. The gears also speed up the motion by going from a large gear on the wheel to a smaller gear on the beaters.

In an electric mixer, the motor does the work of turning the beaters. But how does one motor connect to two beaters? And if the motor's drive shaft is horizontal, how does it spin

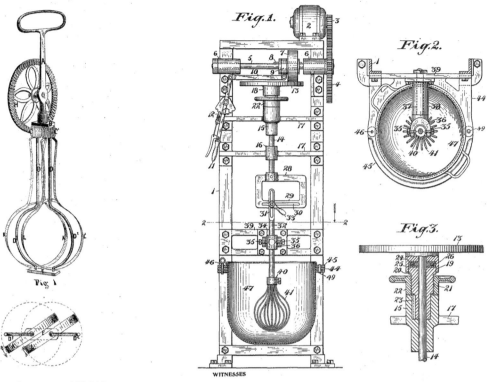

Patent no. 103,811

Patent no. 917,289

two vertical beaters? A worm gear satisfies both these conundrums. This is a gear that has its grooves cut at an angle to the axis of the shaft. Worm gears greatly *reduce* the speed of rotation, from the high speed of the electric motor to the slower speeds needed in mixing, and they also change the direction of rotation 90 degrees. The worm gear is attached to the motor's drive shaft, and it turns two counter-rotating gears attached to the sockets that hold the beaters. Having the beaters turn in opposite directions draws the food mixture inward toward the beaters to mix it and then forces it outward.

Inside the Mixer

The mixer shown here had done some serious mixing in its life. Rescued from the Goodwill Outlet in Seattle, it was seriously grimy. Removing four Phillips screws allowed the top housing to come off, revealing all the inner workings.

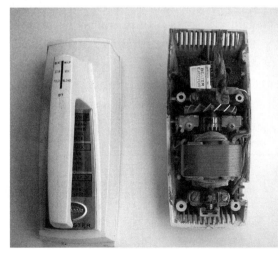

Projecting upward through the top of the case is the switch that controls the speed of the motor. When the switch is moved to a different one of the speed options visible on the upper housing, it connects power to a different wire in the motor.

At the front of the motor assembly are the two plastic gears that drive the beaters. They sit on opposite sides of the worm gear mounted on the motor shaft, so as the worm gear turns, it drives one clockwise and the other counterclockwise. The motor shaft also turns a fan that moves air through the housing to cool the motor; in this mixer it's on the opposite end of the motor shaft, though in other models its may be directly above the worm gear.

Behind the motor is a bearing that holds the motor shaft in place as you push the beaters against the sides of the mixing bowl, while still allowing the shaft to turn. Adjacent to the bearing are two metal prongs that provide a connection for the electric power cord.

Behind one of the plastic gears is a triangular metal arm, which serves as the beater ejector. Depressing the plastic button that sits atop the arm pushes down a metal plate that pushes the beaters out of their sockets.

OVEN (ELECTRIC)

History of the Electric Oven

Home kitchens had ovens starting about 5,000 years ago. A bit more recently, the ancient Greeks developed the art of making bread and pastry in ovens. In 1892 an electric oven was developed by Thomas Ahearn, an inventive Canadian. (That same year he also invented an electric water heater, electric space heater, and electric iron.) Ahearn's oven was made of brick and had peep holes so that the cook could watch the food inside.

Self-cleaning ovens were invented by engineers at General Electric in 1964. They were awarded patent number 3,125,659. The breakthrough recognized by this patent is the self-

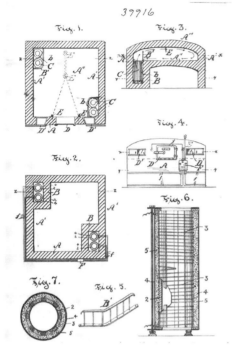

Canadian patent no. 39,916

interrupting thermal relay, which controls heat by controlling the electric power supplied to the oven, which can differ in voltage from one house to the next. This discrepancy can cause great variation in the temperatures of different ovens, but the thermal relay compensates by distributing power to different heating elements depending on the voltage. It allows the self-cleaning oven to operate within its design temperatures.

Because accidentally opening a self-cleaning oven's door during the cleaning process could cause serious burns, in the 1960s inventors also developed a special lock that seals the door shut whenever the temperature inside is sufficiently high.

The first convection oven designed for kitchen use, called an "air-conditioned electric cooking oven," was invented in 1946 by Herbert Mills (patent number 2,408,331). (In truth, all ovens are convection ovens, since they rely on convection currents of warm air to heat the food. But whereas in conventional ovens these currents all occur naturally as air warms and rises, then cools and sinks, convection ovens use a fan to force additional air motion.) Mills was a design engineer who held more than 40 patents; his early work included inventing the window in the oven door. He launched his own design firm, Mills Engineering Company, and later Mills Products, to manufacture some of his inventions. Before Mills made his air-conditioned kitchen oven, others had used forced convection ovens for industrial uses.

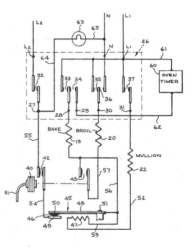

Patent no. 3,125,659

How Electric Ovens Work

Ovens operate in two or three modes: baking, broiling, and, in many ovens today, a mode for self-cleaning. In baking, a heating element at the bottom of the oven radiates heat throughout the oven, causing the temperature of the oven walls to rise. (For more information on how heating elements work, see the introduction, p. xiii.) Air in the oven comes in contact with the walls, which raises its temperature as well. The warm air surrounds the food, cooking it. A thermostat turns the heating element on and off to maintain a constant temperature.

A steak calls for a different form of cooking: broiling. Here a heating element in the *top* of the oven comes on, and stays on until you shut it off. The heat radiates from the heating element to cook the meat below.

In self-cleaning mode, the oven is heated to an especially high temperature (750 to 950° F) that burns off all the spilled lasagna and cookie dough so you don't have to scrub it away. A safety lock keeps the door closed as long as the interior temperature is 600° F or higher. After the oven cools down, which takes a couple of hours, check for ashes. Wipe out any you find. If there are none, and if your oven allows you to choose the cleaning time, reduce the time when you next clean the oven. This will save some energy.

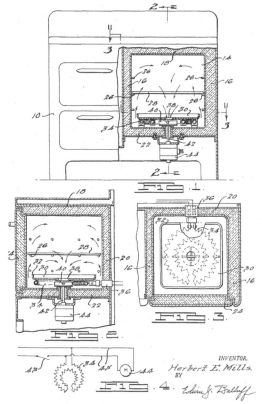

Patent no. 2,408,331

Convection ovens operate like conventional ovens, with one difference: the fan that forces the air inside to move. Heating elements still heat up the air, but the fan blows air cooled by contact with the cooking food away so hotter air can take its place. The constant circulation reduces cooking times by about 20 percent or reduces the required cooking temperatures by 50° F. To prevent the fan from being damaged by the oven's high temperatures, it is housed in a separate compartment.

Inside the Electric Oven

The action is mostly at the back of the oven. Removing the panels reveals the circuit board and switches—many of the operations of today's ovens are controlled by electronics. Electric ovens have

visible heating coils: the one at the bottom for baking and the one at the top for broiling. In many ovens the temperature sensor is visible sticking out of the back wall. It has a metal sheath and is about the size of a small pencil.

In a convection oven, a fan is mounted in the back wall. You can see it as soon as you open the oven door. Even if the oven itself is gas, the fan runs on the house's electric current.

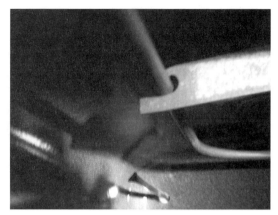

Temperature sensor

Convection fan

Oven (Gas)

History of the Gas Oven

Gas ovens evolved from the cast-iron stoves that were invented in the mid-19th century. Inventors such as Robert Bunsen, who developed the Bunsen burner in 1885, demonstrated that gas could be used effectively for heating as well as lighting, but gas stoves didn't become popular until a system of pipelines was installed to distribute natural gas to homes, beginning in the late 1800s and accelerating after World War II. In 1915 thermostat-controlled gas ovens hit the market. These ovens were able to maintain a constant temperature, which allowed for more even and precise cooking.

Direct ignition of gas flames at the burner became possible when inventors began to develop electronic igniters. In 1976, Elmer Carlson invented an electronic ignition system that not only provided the spark but also controlled the valve that allowed gas to flow (patent number 3,938,938).

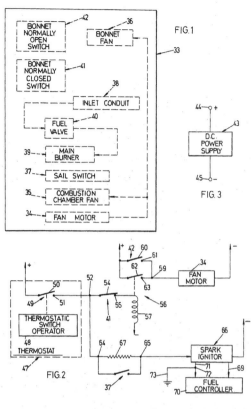

FIG.1

FIG.3

FIG.2

Patent no. 3,938,938

How Gas Ovens Work

A gas oven derives its heat from burning gas instead of supplying electricity to heating coils, but otherwise it works in much the same way as an electric oven. See the chapter on electric ovens (p. 111) for information on the device's basic functions.

Gas ovens ignite in one of two ways. The first is via pilot light. The pilot light is a small flame that is constantly burning inside the oven's burner assembly. If your oven has a pilot light, you can see the flame when you peer inside the broiler drawer. When you turn on the oven, the pilot flame gets bigger. This heats a temperature sensor called a thermocouple, which opens up the gas to the oven's burner. It can take as long as 90 seconds after you turn the dial. You can hear the gas ignite.

If your oven makes a clicking sound when you first turn it on, it uses the second method: an electronic igniter. An electronic device called an oscillator generates an electric pulse that causes sparks between two electrodes to ignite the gas. It will continue to spark for a few seconds or until the gas ignites. A microprocessor opens the gas valve and powers the sparker, and a sensor monitors the flame and shuts off the gas valve if the flame goes out.

Newer electronic igniters incorporate another innovation: the automatic reignition system. When you turn the dial to the desired level of heating, this system senses whether the burner flame is lit; if not, it starts the electronic igniter. Once the flame is burning, it automatically turns off the igniter. In the unlikely event that the flame goes out even though the gas is still flowing, the reignition system will restart the igniter. This magic is performed by two electrodes that sit in the path of the gas and sense its electrical conductivity. Electricity can jump from one electrode to the other only when the flame is burning gas, because the combustion releases free electrons that can carry electricity. If the flame is extinguished, the flow of electricity is interrupted, which triggers the igniter to restart.

Regardless of the ignition system, a gas oven's main burner is located at the bottom of the oven. Some gas ovens have a top burner for broiling, which will have its own igniter. Less expensive models use the bottom burner for both broiling and baking; the broiling is done in a separate tray below the burner.

Inside the Gas Oven

A gas oven's burner sits beneath a metal pan called the flame spreader, which distributes the heat over a large surface so it can heat the air inside the oven. This in turn sits below the oven bottom, which in most models is removable. Adjacent to the burner is the ignition device, whether pilot flame or electronic igniter.

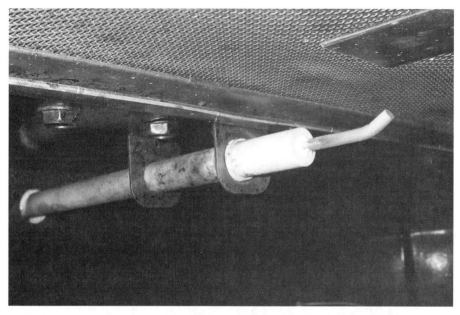

Igniter

Gas ovens often have gas expansion thermostats rather than bimetallic thermostats. With the gas expansion thermostat, a bulb filled with an inert gas is mounted at the top of oven. As the inert gas heats up, the bulb expands and presses against a diaphragm. The diaphragm pushes on electronic controls, which decreases the flow of fuel to the burner. When the inert gas cools and the bulb contracts, the flow of fuel increases.

Thermostat housing

PEPPER MILL

History of the Pepper Mill

Pepper has been used as a spice since before the dawn of recorded history. Millennia later, in the 15th century C.E., Europeans' desire to find less expensive sources for pepper and other spices drove world exploration. They mounted expeditions down along the west coast of Africa and ultimately into the Indian Ocean. Columbus planned to reduce the sailing time to the Spice Islands in Indonesia by heading west rather than around Africa, which is how he discovered the New World.

Before pepper mills, people ground peppercorns by smashing them with a mortar and pestle. In the United States the first pepper-mill patent went to a Frenchman, Pierre Cha-

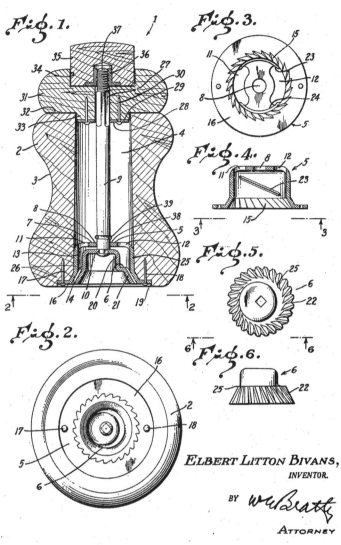

Fig. 1. Fig. 2. Fig. 3. Fig. 4. Fig. 5. Fig. 6.

ELBERT LITTON BIVANS,
INVENTOR.

BY

ATTORNEY

Patent no. 2,407,808

las, in 1878 (patent number 210,837). The design of the pepper mill disassembled in this chapter was patented in 1946 (patent number 2,407,808).

How Pepper Mills Work

Peppercorns are the dried fruit of the black pepper plant. To release their flavor, they are ground into tiny flakes of pepper between two rough surfaces. The resulting flakes provide a much larger surface area for your taste buds to enjoy, which accentuates their spiciness compared to eating the peppercorns whole. But why grind pepper yourself rather

To gain some attention at dessert time, grind some pepper onto your vanilla ice cream. Before your friends have you hauled away, have them taste it. Pepper enhances the vanilla flavor. Really.

than buying it already ground? Pepper, like coffee, loses flavor when the fruit is exposed to light and air. The expanded surface area of ground pepper means that it loses flavor much more rapidly than whole peppercorns.

A pepper mill allows you to grind pepper finely by twisting a knob or handle, crushing the seeds against a cutting surface inside. Some mills use an alternative approach: rather than rotating the mill, you squeeze two handles together, which operates a lever and forces the seeds past grinding plates. A spring forces the handles open again.

Be sure to read the label on any pepper mill you purchase to determine whether it can also be used to mill salt and other seasonings. Most pepper mills have steel or zinc-alloy grinding burrs that can corrode if used for milling salt. Mills that use ceramic or plastic burrs can mill both salt and pepper. Dedicated salt mills generally have stainless steel burrs.

Inside the Pepper Mill

The nut on top of the pepper mill unscrews and releases the knob from the grinding shaft, which allows you to fill the base of the mill with peppercorns. The only thing holding the shaft inside the base is the plastic dial at the bottom that changes the coarseness of the grind. The dial unscrews, and this allows the shaft to fall out.

Attached to the bottom of the shaft is the cutting surface; a ridged ring remains attached to the base of the mill. When the mill is assembled, twisting the knob turns the cutting surface, grinding peppercorns between it and the surface of the ring. Tightening the dial raises the cutting surface up, closer to the ring. This allows only finer grains of pepper to fall through.

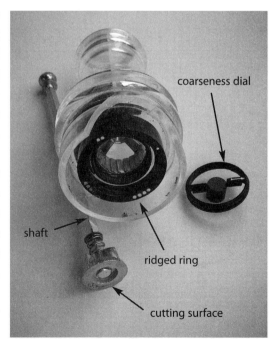

coarseness dial

shaft

ridged ring

cutting surface

POPCORN POPPER

History of the Popcorn Popper

Native Americans shared their secret popping corn with the Spanish conquistadores, who took some back to Europe. Originally, popcorn was prepared over an open-hearth fire; people have been making popping baskets for that purpose for many years. But with the rise of the industrial age, popcorn lovers began inventing a variety of new machines to keep the corn popping.

Charles Cretors was the Edison of roasting machines. He held many patents for popcorn poppers and peanut roasters. In 1893 he patented a machine that heated the popcorn—or coffee, or peanuts—with steam (patent number 506,207). It was a 400-pound device that was supposed to be portable "by a boy or by a small pony," according to a *Scientific American* article of the day. One of Cretors's later inventions was an electric popper, patented in 1922 (patent number 1,436,862).

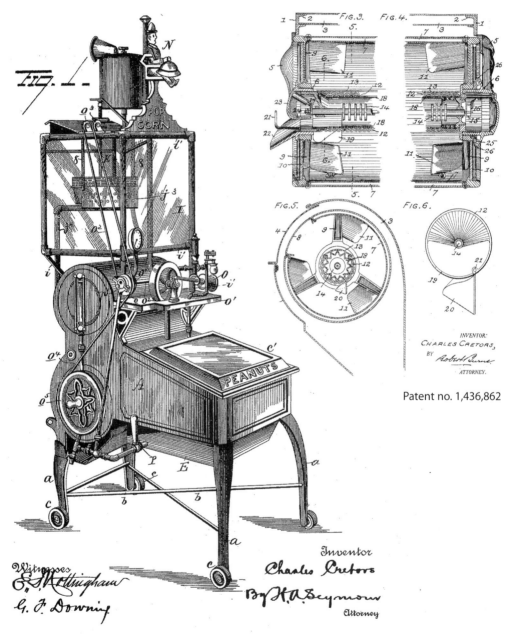

Patent no. 1,436,862

Patent no. 506,207

Cretors started out in the retail candy business. When a peanut roaster he purchased proved disappointing, he decided to improve it. A traveling salesman told Cretors that he could sell the roaster that Cretors had made. That combination of talents launched C. Cretors and Company, which still manufactures a great variety of popcorn poppers, mostly for movie theaters and carnivals.

How Popcorn Poppers Work

Popcorn is a type of corn that contains a large amount of water (about 14 percent by weight), which is sealed inside each kernel's hard shell, or pericarp. When the kernels are heated to 400° F, the water vaporizes. The volume of the water vapor is many times greater than the volume of the liquid, which greatly increases the fluid pressure inside the pericarp. When the pressure is about eight times that of the atmosphere outside, the shell explodes, which causes the audible "pop." The water vapor also inflates the starchy material inside the kernel, so that after it has popped, the corn's interior is much larger than its outer shell. The kernel has essentially turned inside out.

The popcorn popper depicted in this chapter is typical of the inexpensive models on the market. It has two components: a heating element and a fan. The fan blows air through the corn, and since the popped kernels have a much greater surface area with which to catch the air, they're carried up and away from the heating element. This makes room for more unpopped kernels to fall down closer to the heat.

Inside the Popcorn Popper

The heating element and fan assembly pull right out from the bottom of the machine when the cover is removed. (For more information on how heating elements work, see the introduction, p. xiii.) Connected to the heating element are both a thermostat and a thermal fuse. The thermostat is a simple bimetallic device, with parts that respond differently to changing temperatures. As the thermostat is heated, one of its arms expands and bends away from the other, breaking

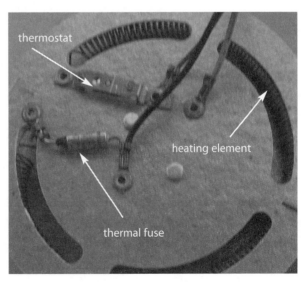

the electrical circuit. With the current removed, the heating element shuts off and the temperature begins to drop. When the thermostat cools, the thermal switch bends back into its previous position, letting current flow to the heating element again.

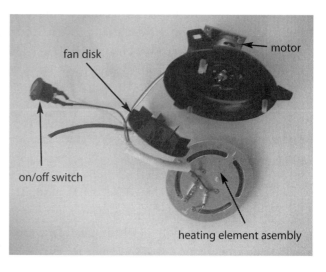

The thermal fuse, meanwhile, contains organic material that melts if temperatures rise beyond safe levels, opening the fuse and stopping the flow of current to the heating element. This is a one-shot device intended only to protect people in case of malfunction. If it breaks, it must be replaced.

The fan in this model is powered by an inexpensive direct current motor. It sits on a circuit board along with four diodes. Diodes pass current only in one direction, rectifying the alternating current from the wall outlet into direct current to power the motor. As the motor spins, it turns a plastic disk that rides on the motor shaft. The fan blades are molded onto the bottom side of the disk. They move air outwards and it escapes through slots in the heating element assembly that sits atop the fan.

PRESSURE COOKER

History of the Pressure Cooker

Denis Papin invented the pressure cooker in 1679. As an assistant to Anglo-Irish physicist Robert Boyle, Papin was familiar with experiments in high pressure. He called his device a "steam digester." His invention was not adopted for home use until much later. Early models had a pressure relief valve, but often it wasn't up to the challenge of reducing the pressure inside a container heated by an open fire. Dinner frequently arrived with a bang, and sometimes people were injured.

A pressure cooker design from 1875 (patent number 170,404) allowed the cook to add water during the cooking process. Carl Nelson invented a new model in 1929 (patent number 1,826,947) and assigned rights to the National Pressure Cooker Company, which demonstrated it at the New York World's Fair in 1939. The National Pressure Cooker Company is now National Presto Industries, one of the leading manufacturers of pressure cookers and other kitchen devices, including the Salad Shooter (see p. 139).

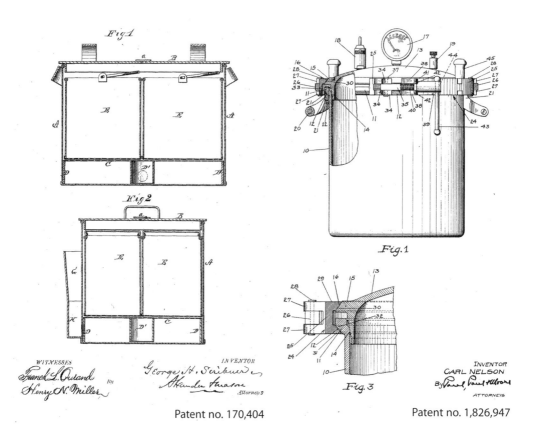

How Pressure Cookers Work

Steam is very effective at transferring heat to food being cooked. This rapid and efficient heat transfer reduces cooking time and nutritional losses in cooking. But to cook food properly, the steam must be at a high enough temperature. This is where pressure cookers can help.

As everyone who lives in Denver knows, at higher elevations—where the atmospheric pressure is lower—water boils at a lower temperature, and foods take longer to cook. Pressure cookers take the pressure effect in the other direction and *raise* the internal pressure. That way, the water inside boils at a higher temperature, and foods cook more quickly. The combination of high pressure and high temperature cooks foods 3 to 10 times faster than cooking them in an unpressurized pot.

Pressure cookers are not all that common in the United States, but they're more popular in the Europe, Asia, and South America.

Inside the Pressure Cooker

A pressure cooker is a pot made of aluminum or stainless steel. To maintain pressure inside, the pot and its lid interlock. A rubber gasket inside the lid forms a seal between them. (Any food particles adhering to the gasket can prevent the top from sealing correctly.) Traditional pressure cookers have a "jiggle top" pressure regulator, a weighted cap that sits on a vertical vent tube in the center of the lid. The weight of the cap keeps the pot sealed, but eventually the pressure builds enough to push the cap out of place. Air and steam rush out of the vent, and as the pressure drops, the cap returns to its original position, resealing the pot. The jiggle top's back-and-forth action releases pressure at a controlled rate.

How hot does it get inside a pressure cooker? Most cookers raise the boiling point of water to about 257° F. They operate at about 2 atmospheres of pressure, about 15 pounds per square inch above sea-level pressure.

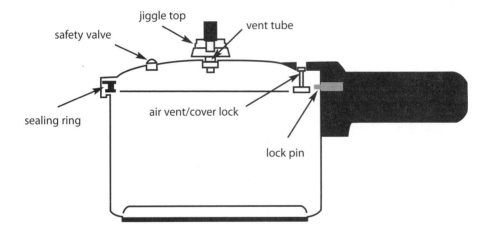

Newer designs have replaced the noisy jiggle top with a spring-regulated pressure dial. With these, you dial in the pressure you want.

Pressure cookers also have a safety valve, a metal pin set inside a rubber gasket. As pressure builds, the pin is forced outward, sealing against the gasket to prevent air from escaping. If, however, pressure rises to unsafe levels, the gasket and pin will be forced out in an explosive reaction (that I'd like to see some day, provided I'm standing in someone else's kitchen). Pressure is released, preventing the pressure cooker from exploding and doing more serious damage to the kitchen and cooks.

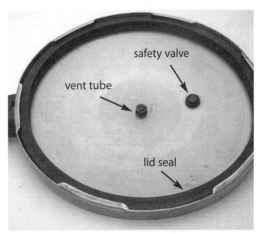

RANGE HOOD

History of the Range Hood

For as long as there have been kitchens, people have needed to vent cooking fumes out of them. Long before electricity was harnessed and charcoal filters were invented, there were fume hoods. It was not until 1865, however, that the first U.S. patent was issued for a hood for cooking stoves (patent number 48,592). The purpose of this invention was "conveying off steam, gas, and odor from the stove to prevent the well-known disagreeable results of having it distributed through the house."

How Range Hoods Work

Range hoods most often take the form of a wall-mounted or ceiling-mounted canopy that hangs over your stovetop. Inside is a motor-driven fan, which withdraws the air and fumes

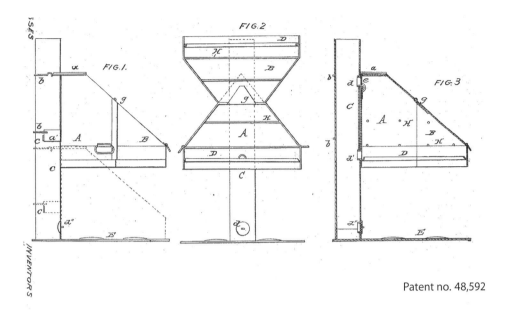

Patent no. 48,592

that rise up from the range top like an olfactory smoke signal. The fan can be either the more typical rotary fan, which is noisier and cheaper, or the larger, more expensive, and quieter centrifugal fan.

The biggest difference in hoods is what happens to the collected fumes. Less expensive models pull the fumes past a filter and circulate them back into the kitchen. More effective are units that push the gases through a vent to the outside. Ideally, the ventilated hood and the range it covers are located on an exterior wall; otherwise, additional ductwork has to be run from the hood to the vent.

For ranges on kitchen islands without wall access, either the hood can be mounted downward from the ceiling, or a pop-up venting system can be installed: push a button and a motor raises a partition behind the range, inside of which is a fan to draw in the fumes.

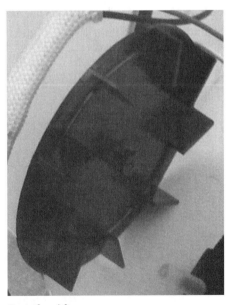

Centrifugal fan

Filters can be made just of aluminum mesh, or they can be filled with charcoal, which makes them more effective at capturing smells. Aluminum filters can be washed periodically in the dishwasher to remove the grease and dirt, but charcoal filters have to be replaced.

Some hoods have heat alarms, and most have lights to help illuminate the omelet you are burning

Inside the Range Hood

Removing the aluminum mesh filters of the hood depicted in the chapter reveals . . . not much at all. Inside is the ductwork that carries air, obnoxious smells, and volatiles away from the stovetop. In this case they all continue up a duct to the outside for venting. Also inside are housings for lights that shine down on the range below. This hood is integrated into a microwave oven above, so the fan is hidden behind the components that operate the oven.

Aluminum mesh filters

Refrigerator

History of the Refrigerator

Not so long ago, before electric refrigerators were invented, people kept food cold in ice-boxes. These double-walled wooden boxes were insulated and lined with zinc or tin sheeting; they included one compartment to hold food and another to house a large block of ice. Ice vendors cut blocks of river and lake ice in the winter and stored them for later sale to consumers. The iceman, in a horse-drawn wagon or later in a truck, delivered a chunk to each home. Heat inside the icebox melted the ice instead of raising the temperature of the food. Drippings from the melting ice fell into a pan on the floor below, which could be removed and emptied.

In 1857, Australian James Harrison invented the vapor-compression method of mechanical refrigeration that we still use today in most refrigerators. Warm winters in the United States in 1889 and 1890 reduced supplies of natural ice and stimulated the adoption of mechanical refrigeration. At first, ice vendors used the technology to manufacture uniform blocks of ice and deliver them to consumers for their iceboxes.

The first refrigerator designed for home use was invented by Frenchman Marcel Audiffren in 1895 (patent number 551,107). It used a vapor-compression design, with sulfur dioxide as the cooling medium. Although it was manufactured and sold in the U.S. starting in 1911, it never caught on, possibly due to its high price—it cost about twice as much as a new car! The first patent by an American for a kitchen refrigerator was issued to William Merralls in 1909 (patent number 927,571); his designed cooled food by blowing air with an electric fan past water-soaked material so the water evaporated. But it was vapor-compression technology that ultimately ended the reign of the ice industry and its teams of door-to-door icemen. A few years would pass before vapor-compression devices replaced the much less expensive iceboxes, but the end of World War II launched a surge in refrigerator purchases.

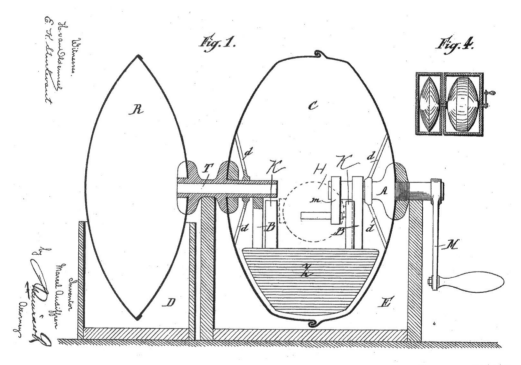

Patent no. 551,107

The refrigerants used in the early vapor-compression refrigerators gave off toxic gases and in a few cases caused deaths. DuPont developed Freon as a safer alternative to earlier refrigerants; it was adopted by refrigerator manufacturers in the 1930s. In the 1980s, Freon was replaced by newer chemicals, because it and other chlorofluorocarbons were destroying the Earth's ozone layer.

There have been many other advances in refrigerator technology over the years, and manufacturers have developed many new features. One of my favorites is automatic defrosting. As a kid I used to help periodically scrape the ice off the sides of the tiny freezing compartment of the fridge. Automatic defrosting eliminated this task. Add an ice maker/water dispenser built into the door, and you have the modern refrigerator.

How Refrigerators Work

Refrigerators keep food cold, but not frozen—at around 40° F. At this temperature, bacteria and mold grow very slowly, so food stays fresh much longer than at higher temperatures.

Refrigerators work by transferring heat from the inside of the device to the outside. That's right—while your refrigerator cools the insulated compartment that stores food, it actually heats up the rest of your kitchen. But what if you open the refrigerator door to release an avalanche of cold air? Believe it or not, this, too, ultimately raises the temperature in the room. The thermostat will sense that the refrigerator

Today nearly every home in America (99.5 percent) has a refrigerator.

has warmed up and instruct the heat transfer mechanism to work even harder. Since the mechanism is not perfectly efficient, it will generate more heat than cooling, and the net effect is that over an hour or so the kitchen will get warmer even if the door stays open.

Refrigerators transfer heat by using a pump, or compressor, to move a low-boiling-point refrigerant between two sets of coils. It is the compressor that you hear start and stop. As the refrigerant passes through the compressor, it is in the form of a gas. The compressor increases the pressure of the vapor, which also increases its temperature. The heated vapor flows into the condenser coils beneath (or in back of) the refrigerator, where it cools, allowing the heat to dissipate into the kitchen air.

Now cooled, the gas condenses into a liquid state and

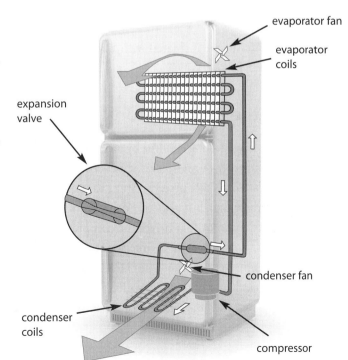

evaporator fan

evaporator coils

expansion valve

condenser fan

condenser coils

compressor

the pressure drops. The liquid goes through a device—an expansion valve—to reduce its pressure even more, which further reduces its temperature. The liquid then flows into the evaporator coils, which are in contact with the air inside the refrigerator. The liquid is much colder (−25° F) than the air in the refrigerator, so it absorbs heat through the evaporator coils. The inside of the refrigerator is cooled and the refrigerant is heated—which returns it to its gaseous state so the cycle can begin again. Through this process, it moves heat from the interior evaporator coils to the exterior condenser coils.

Old refrigerators had separate compartments inside the refrigerator for items to be frozen. The cooling coils surrounded this compartment, freezing its contents and cooling the rest of the refrigerator to a lesser degree. Newer refrigerators have separate freezer compartments with their own coils and thermostats.

Next to the evaporator coils, frost-free refrigerator/freezers have heating elements connected to a timer. Every six hours, the timer switches them on, which melts and evaporates any frost buildup in the freezer. A thermostat switches off the heating elements when the temperature rises to the freezing point. When they switch off, the cooling cycle starts again to maintain the temperature inside.

The most common cause of problems with refrigerators is that their external heat exchange coils get buried in dust and dirt or their internal coils get covered with ice. Outside the refrigerator box, as the detritus of the kitchen builds up around them, the external coils gain a layer of insulation, which slows the exchange of heat. Search for the ubiquitous dust bunnies that hide beneath and behind the refrigerator and suck them out with a vacuum. Keeping the coils clean allows the refrigerator to run more efficiently, saving you electrical power and money. Inside the refrigerator, the cooling coils will work less efficiently if they are insulated with a layer of ice. Replacing a faulty thermostat will often fix this problem.

Inside the Refrigerator

If you're following along at home by looking at the inner workings of your refrigerator, **beware of the electrical hazard.** Of course you want to unplug the refrigerator first, but also know that in order to provide the motor that drives the compressor with higher torque when it first starts up, the motor is connected to a capacitor. The capacitor stores electrical energy, and it can give you a serious shock even when the refrigerator is unplugged.

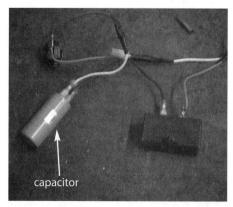

capacitor

The capacitor is mounted on top of the motor in a metal case. If you are going to do anything more than look inside the back of the refrigerator, discharge the capacitor by connecting each of its two legs to a 2-watt, high resistance (20,000 ohm) resistor. Electricity stored in the capacitor will flow through the resistor, heating it and discharging the capacitor.

Along the back of the refrigerator, you may find the condenser coils. If you don't see coils on the back, they are instead located at the bottom of the fridge, where a fan pulls in air from the kitchen and pushes it past the coils, which are wrapped around the fan. The compressor that pumps refrigerant into the condenser coils is also located at the bottom rear of the refrigerator.

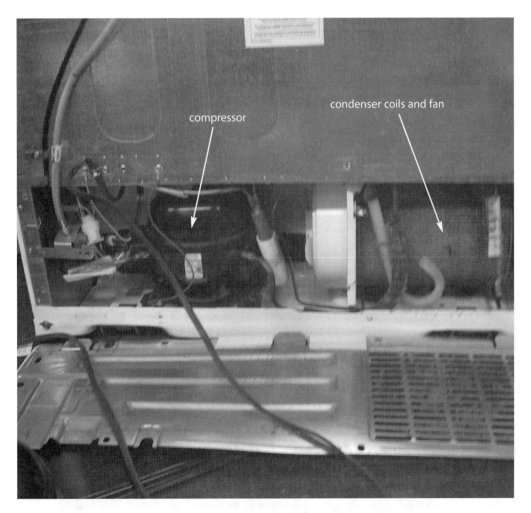

Inside the refrigerator, behind the freezer compartment, are the evaporator coils. A fan adjacent to the coils circulates air over them to ensure a good exchange of heat.

To keep the evaporator coils from becoming covered with ice and losing their capacity to transfer heat, automatic defrosters are used. The defroster has two main components in addition to the heating coil: a timer and a thermostat. Most use gas thermostats, as in the photo. Unlike bimetallic thermostats, which rely on the expansion of two different metals to respond to changes in temperature, gas thermostats rely on the volume change experienced by a gaseous substance as it is heated or cooled.

Electronic controls are housed in a separate compartment at the top of the refrigerator, along the back.

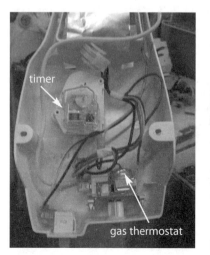

Refrigerator controls

REFRIGERATOR MAGNET

History of the Refrigerator Magnet

Max Baermann of Germany invented the flexible magnet in 1960 (patent number 2,959,832). It's not clear if he envisioned that refrigerators around the world would one day be covered with advertisements and family photographs—his patent covers the use of his invention as a locking mechanism or a clothing fastener, not a refrigerator accessory.

How Refrigerator Magnets Work

A flexible refrigerator magnet is a "one-sided flux" magnet, also known as a Halbach array. To create one, grains of ground magnetic material (ferrite or iron) are spread over a plastic base. The grains are laid in rows, with the nonmagnetic plastic separating each row.

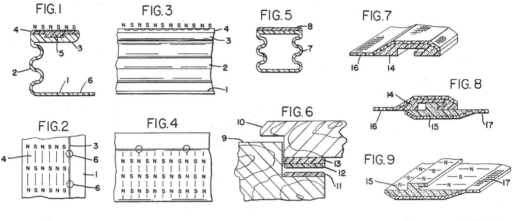

Patent no. 2,959,832

The rows are magnetized into alternating north and south poles. Figure 4 from the patent application above depicts this arrangement.

The resulting magnets are one-sided—they have twice as much magnetic attraction on one face and almost no attraction on the other. Don't believe it? Try putting a flexible magnet on the refrigerator backwards. Does it stick?

Inside the Refrigerator Magnet

If you could look inside the magnet, you would find powered iron or ferrite incorporated into the plastic or rubber.

Test the Fields

Take two refrigerator magnets and place them together with their magnetic sides facing each other, so that they're firmly attached. If you pull one magnet past the other along the grain, the magnets will slide easily past each other. If you pull it in the perpendicular direction, they will alternatively attract and repel each other— you will have to pull hard to get them to move, and then they will slip to the next position as the magnetic rows align.

SALAD SHOOTER

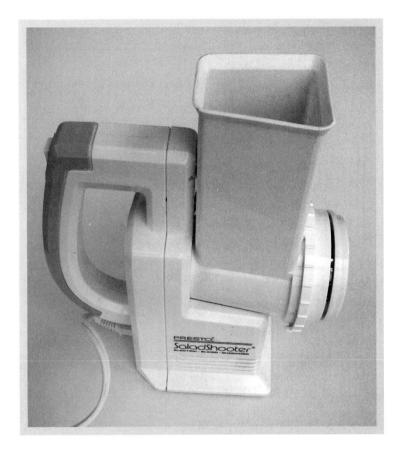

History of the Salad Shooter

The Salad Shooter is the brand name of an electric slicer and shredder manufactured by National Presto Industries. The patent for the model disassembled in this chapter was issued in 1989 (patent number 4,884,755). Presto also obtained a design patent (patent number D300,400) for the device in the same year; while the standard utility patent protects their mechanical design, the design patent protects the external or aesthetic design. Sales of the Salad Shooter started the year before the patents were issued—a typical lag in the world of patents. In 1990 the success of the device led Presto to introduce the Professional Salad Shooter, a beefed-up version of the original.

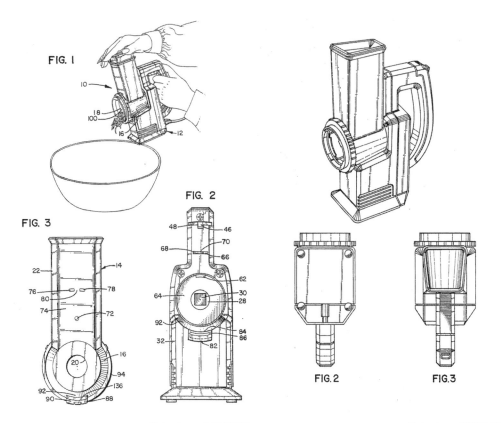

FIG. I

FIG. 3

FIG. 2

FIG. 2

FIG. 3

Patent no. 4,884,755

Patent no. D300,400

How Salad Shooters Work

The Salad Shooter is essentially a motor-driven food processor (see p. 69 for more information on food processors in general). However, whereas other food processors must incorporate a mixing bowl to hold the food being cut, and a large cutting blade that spins horizontally within that bowl, the Salad Shooter usually has a smaller blade that spins vertically, embedded in an open-ended cone that allows the cut food to spill out into a separate container. As a result, while most food processors are too heavy to lift off the countertop while in use, the Salad Shooter is light enough to pick up and hold during operation—you can aim it to shoot carrots at your friends.

When you turn on the Salad Shooter, the motor spins, and after the rotational speed is reduced by gearing, it turns the processor cone with its cutting or shredding blades. You drop foodstuffs into the chute on top and they fall into the spinning processor, which slices them into slivers and chunks as they pass into the interior of the cone. The pieces then fall out of the wide open end of the cone—hopefully into a waiting bowl.

Inside the Salad Shooter

The two principal parts disassemble with a twist. One half holds motors, gears, and switches, and never touches the food. The other half consists of the chute the veggies plunge down and the whirling processor blades that cut them.

On the back of the chute is a latch that fits into a slot in the motor housing. If this latch isn't seated correctly, the motor won't spin. It's a nice safety feature. But there's more—a second safety latch farther down closes only when the processor assembly has been rotated into the correct position for firing. So even if some mischievous kid figures out how to defeat the first safety latch, there is a second one he or she has to beat, too.

Next, detach the processor cone from the chute. The cone is plastic and has metal blades or shredders protruding from the side. The piece that holds the metal cutters can spin inside its housing, and at the narrow end of this piece is a large plastic notch. It fits into a square-shaped slot on the motor housing, from which it receives the mechanical power to turn.

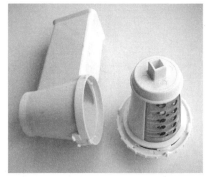

Chute (left) and processor cone (right)

Removing a few screws opens up the motor housing. Two large plastic gears convert the high-speed output of the motor gear into a more reasonable, slower speed suitable for shredding. Counting the teeth on the three gears, I calculate that the rotational speed of the cutter drops to about $1/81$ of the motor speed. The motor turns at 3,600 RPM, so at the end of the gear reduction the processor cone turns at about 44 RPM, less than once per second.

Digging deeper, you can extract the motor. On the opposite end of the motor shaft from the gearing is a plastic fan to circulate cooling air onto the motor. Electrical components include the connections that enable the safety latches, and a thermal fuse that will break the circuit if the motor gets excessively hot.

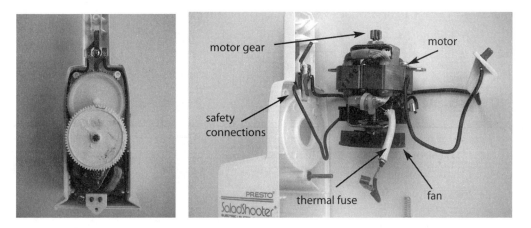

motor gear

motor

safety connections

thermal fuse

fan

SALAD SPINNER

History of the Salad Spinner

Gilberte Fouineteau, a French inventor, developed the "domestic appliance for drying vegetables" in 1975 (patent number 3,885,321). His patent claims rights to a rotating basket, turned by a hand crank, that "may be used for the draining of salads, having a removable basket useable independently with an inner volume that is completely free."

How Salad Spinners Work

Centrifugal acceleration! Remember the Rotor ride at the amusement park? You stood inside a giant spinning tub, and as the tub spun faster (up to 30 revolutions per minute), the floor that you walked on to enter dropped away. But instead of falling, you were pinned against the outer wall. In the same way, as the salad spinner turns around and around, the

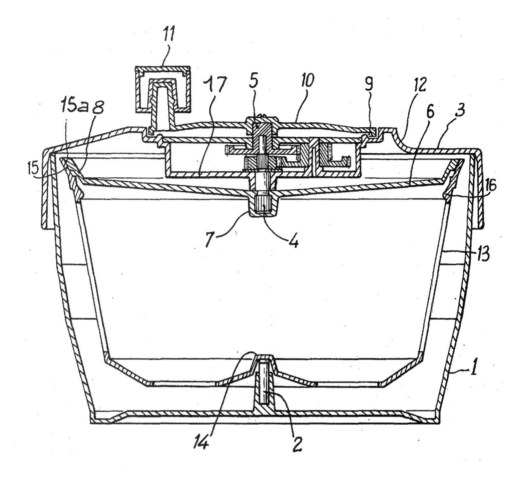

Patent no. 3,885,321

lettuce is thrown outward, against the sides of the basket. Unlike the amusement ride's walls, which are solid, the salad spinner's basket has many slots. So as lettuce is forced against the basket walls, water is forced *through* them—it slides through the holes in the inner basket and into the outer bowl.

A salad spinner may use many different methods to get the inner basket spinning. A trip to an upscale kitchen supply store showcases all the creative ways. Simplest is the hand-cranked design invented by Fouineteau, by which the user spins a handle that turns gears that spin the basket.

Inside the Salad Spinner

The salad spinner disassembled in this section, however, is not the hand-cranked type, but a later design that features push-to-spin operation. Some kitchen devices are elegantly

Old-Fashioned Salad Spinning

I'm tempted to try this, but I'll let you do it instead. Wrap up washed veggies in a clean dish towel. Grab both ends of the towel and twirl it around your head in big circles. After a few seconds of vigorous lassoing, the towel will be wet and the veggies dry. Just don't let go of one end of the towel while spinning.

designed to allow for effortless reverse engineering; loosen a few screws and the innards pop out. But not this spinner—it required some wrestling.

To operate the spinner, you release the pump handle from the lid by sliding back a latch on the lid. This allows a spring surrounding the handle shaft to push the handle up. Inside the handle is a metal screw, about the size and appearance of a corkscrew. Pushing the handle back down forces the screw through a plastic spinner piece that is latched to the inner basket.

Push-to-spin salad spinner

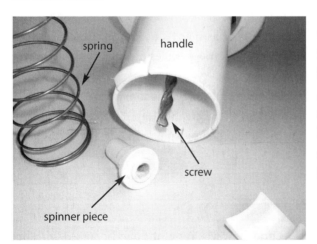

spring

handle

screw

spinner piece

As the handle pushes the screw through the spinner piece, the piece rotates, spinning the inner basket inside the stationary outer bowl. After every pump of the handle, the spring pushes it back up so it can be depressed again and the spinning can continue until the lettuce inside is dry.

SARAN WRAP

History of Saran Wrap

The group of plastics that are used in Saran Wrap were discovered by accident in 1933 by a Dow Chemical employee, Ralph Wiley. At the time Wiley was not a research chemist but a college student who was hired to clean glassware in the lab. One day while scrubbing some glassware used in an experiment with dry cleaning products, he found himself unable to remove a sticky plastic material. He quickly found that the residue had its own applications. Although it had a green tint and an offensive odor, it could be used to protect army boots, airplane wings and fuselages, and car upholstery.

In 1933, Dow Chemical was awarded the first patent that described how to make these new chemicals, called carbon chlorines (patent number 2,034,292). The patent named Wiley as one of inventors, and he would be recognized again in many subsequent chemical patents.

Dow chemists later discovered how to produce this material without the odor or green tint, which allowed the company to introduce Saran Wrap for industrial use in 1949 and

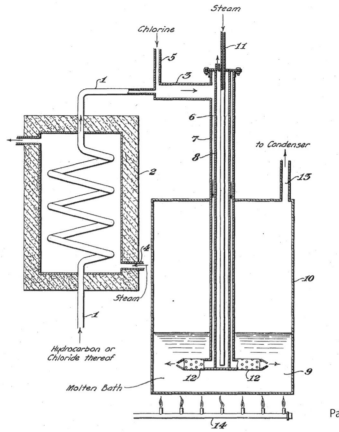

Steam

Chlorine

11

5

3

1

6

7

to Condenser

8

2

13

4

10

Steam

1

9

Hydrocarbon or
Chloride thereof

12 12

Molten Bath

14

Patent no. 2,034,292

for use in the home in 1953. Although the clingy plastic wrap is the best known application for Saran, most of the Saran produced is actually used as a liner in other packaging.

The product's chemical formula was changed in 2004; it now uses low-density polyethylene to reduce environmental concerns about the chlorine in the plastic.

How Saran Wrap Works

The value of Saran Wrap is that it is largely impermeable to water, oxygen, and odors. Food wrapped in Saran Wrap retains its moisture, doesn't get exposed to air and spoil, and doesn't stink up the refrigerator. The secret to its success is that it is made up of long molecules—polymers—that bind so tightly to each other that water and other molecules can't pass between them.

And why does it cling? Electrostatic forces! As the plastic is made, it is rolled and unrolled, and in the process it rubs against itself and picks up static electrical charges, like rubbing your feet on the carpet. And when you wrap a piece over a salad bowl, you cause some electrical charge to move between the bowl and wrap. This leaves areas of positive and negative charges that cling to each other.

SMOKE DETECTOR

History of the Smoke Detector

The first portable electric fire alarm was invented in 1890 by Francis Upton—an associate of Thomas Edison's—and Fernando Dibble (patent number 436,961). But most of today's home smoke alarms employ one of two more recent detection systems. The first is the optical smoke detector, patented in 1942 (patent number 2,298,757). The second, the ionization smoke detector, was patented in 1967 (patent number 3,353,170). A few years later, in 1973, Duane D. Pearsall was awarded a design patent (patent number D226,539) for a home unit that is credited with launching the home smoke detector business. Now, of course, smoke detectors are a standard feature in almost every home.

How Smoke Detectors Work

Smoke detectors have two component systems: a detector and an alarm. In optical smoke detectors, the first component consists of a light source, usually an LED, and a light

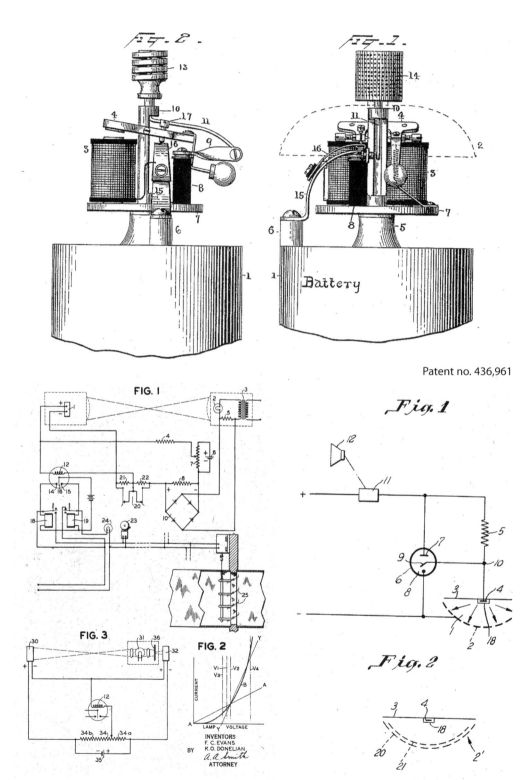

Patent no. 436,961

Patent no. 2,298,757

Patent no. 3,353,170

sensor, or photodiode. The photodiode and LED are not aligned, so when no smoke is present, the photodiode can't see the light. Smoke, however, scatters some of the light into the photodiode, which sounds an obnoxious alarm.

In an ionization smoke detector, on the other hand, the detector component holds a tiny piece of the radioactive element americium. The americium's radiation causes atoms of gas in the air (primarily nitrogen and oxygen) to add or lose an electron, creating positive and negative ions. The ions permit a small current, usually powered by a 9-volt battery, to flow through the air between two charge plates. When smoke particles are present, they attach to the ions and reduce the current flow between the two plates. Sensing a decreased current, the detector switches on that horrible buzzer.

You should replace the batteries in your smoke detector when daylight saving time switches to standard time, and when it switches back again.

The alarm component is a piezo speaker. It contains a crystal that vibrates and produces sound in response to changes in the electrical voltage applied to it.

Inside the Smoke Detector

Start by opening the battery case. Here resides a 9-volt battery. This detector is also powered by house current; the battery is a backup in case power fails. (Most detectors are not powered by house current—the ones that are twist off their ceiling mounts to reveal two wires leading up into the ceiling.) In the center of the detector is a silver-colored piece of metal that acts as a switch. When you press the plastic button on the cover to test the smoke detector, the button depresses this piece, which makes contact, completing the circuit to directly trigger the alarm. That's annoying!

Since this model is an ionization detector, it features a cylindrical ionization chamber, which is controlled by the electronics below it.

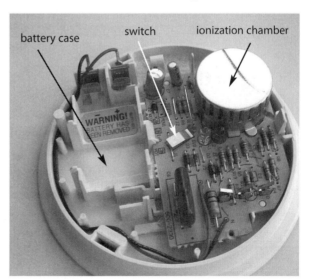

battery case   switch   ionization chamber

STEAMER

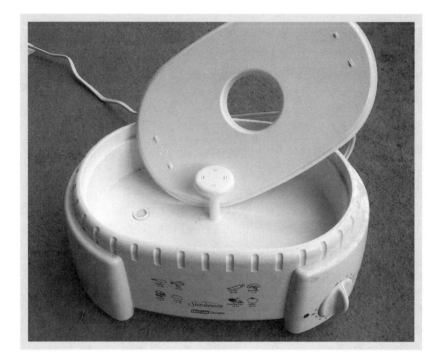

History of the Steamer

Steaming as a cooking process dates back thousands of years. Residents of Asia and Africa steamed rice and other vegetables in baskets made of wood—bamboo, for example. In the West, however, steaming has been slow to catch on. But as people discover the nutritional benefits of steaming, the process is gaining popularity.

Metal baskets that fit inside pots are a simple way to steam vegetables. They are, in essence, an updated version of the wood steamers used for centuries. More recently, inventors have developed electric steamers. For example, five inventors working for appliance manufacturer Rival patented their design in 1985 (patent number 4,509,412).

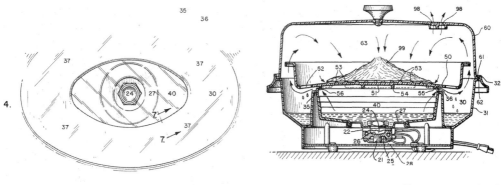

Patent no. 4,509,412

How Steamers Work

As steam condenses back into water, it gives off a tremendous amount of heat. Steamers produce rising clouds of steam to cook food. Fewer nutrients are lost in steaming than in boiling or any other method of cooking, so people concerned with healthy eating often choose to steam their vegetables. Of the two methods of steaming, the simpler is to boil water in a pan on the stove and set a metal basket with holes in it—a colander—in the pan. Vegetables ride in the colander and cook in the steam.

Self-contained steamers, on the other hand, do not require a stovetop and can be plugged in anywhere. They have one compartment for the water/steam and one for the veggies (or hot dogs) being cooked. You add water to the reservoir in the base of the steamer, place the food to be cooked in the chamber above, and twist the dial to start the heater. Water flows into a heating tube, where it warms up and eventually gets to the boiling point. As the water converts to steam, its volume increases greatly, and it rises up the heating tube into a sprayer in the food compartment.

As the steam lands on the food, it condenses into water, transferring its heat and cooking the food. The water drains back into the reservoir and then flows back into the heating tube.

The cooking process is monitored three ways. First, you set a timer. When the timer has finished, it cuts power to the steamer. Second, a thermostat maintains the temperature in the heating tube and cycles on and off with an audible click. Finally, as a safety feature, the steamer has a thermal fuse that interrupts power and won't reset; this only occurs if the steamer has a fault that requires repair.

Inside the Steamer

A clear plastic lid sits on top of the steamer to hold in the steam and hot air. Removing the lid exposes the plastic tray on which the veggies sit. The tray lifts out to reveal the water

Raw > Steaming > Boiling

Any form of cooking reduces the nutrition in vegetables. Steaming vegetables reduces their nutrition less than other methods, including boiling.

reservoir in the steamer's base. A few screws secure the plastic cover to the base, and once the cover is removed, the heating system is visible on the underside of the reservoir.

The steamer's heating system looks remarkably like that of an electric coffee percolator (see p. 21). Water flows through a curved heating tube that lies next to the heating coil; both are encased in the metal sheath that is visible in the photo. (For more information on how heating elements work, see the introduction, p. xiii.) Water flows down into the

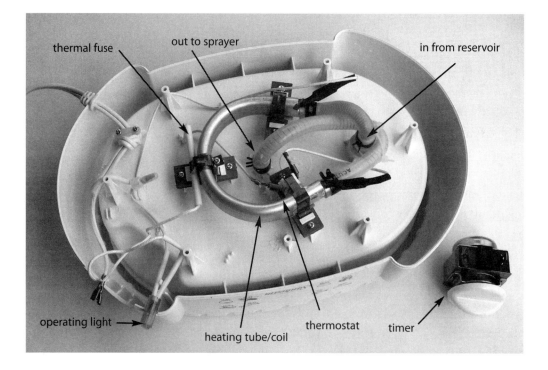

tube from the reservoir above and leaves the tube to spurt out the top as steam and boiling water, then rain down on the broccoli below. On the side of the heating tube/coil is a straight, white section of tubing that holds the thermal fuse. Similarly, the thermostat is glued to the side of the heating tube/coil to sense accurately the temperature of the tube. Electrically, the heating element is connected in series not only with the thermostat and thermal fuse but also with an operating light.

The steamer's power switch is on a timer. You twist the dial to start the timer, which is a spring-powered mechanical device. When time runs out, the slot in the black wheel on the back of the timer rotates down so the spring-powered lever beneath it can jump upward. When this happens, power to the steamer is cut.

Timer mechanism

STOVE

History of the Stove

In 1802, George Bodley invented the first cast-iron cooking stove with a flue to vent the gases produced during combustion. The cast-iron stove allowed cooks to bake, grill meat, and cook on the stovetop all at the same time. Wood or coal fueled the stove. Bodley had started a foundry in 1790 in Exeter, England, and the stove he designed became a major source of business for his foundry. It stayed in business for 177 years.

The first American stove inventor was James Greer, who in 1843 was awarded patent number 3,084 for his cooking stove. Greer continued tinkering with stoves and won several more patents.

Gas as a fuel for cooking was first demonstrated in 1802 by Frederick Albert Winsor. However, gas stoves were slow to catch on. Not until gas pipelines were built (starting in the late 19th century and accelerating after World War II) to deliver gas to homes did they

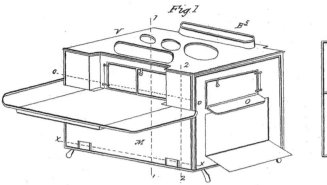

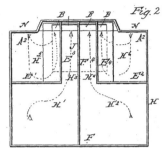

Patent no. 3,084

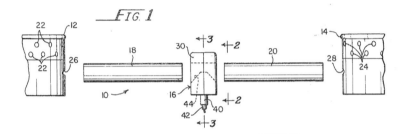

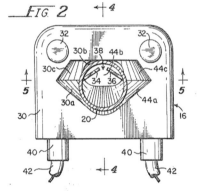

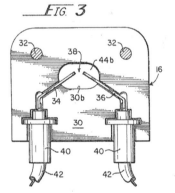

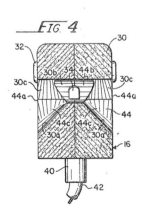

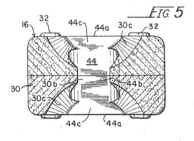

INVENTORS:
GILBERT W. GROSS
LEONARD H. MICHAELS
WILLIAM H. PATRICK

By Mason, Kolehmainen
Rathburn & Wyss

Patent no. 3,523,747

finally become popular. A parallel development that contributed to gas stoves' popularity was the electronic gas ignition system, which pushed aside the pilot light and made the devices safer and easier to use. Gilbert Gross and several other inventors were awarded patent number 3,523,747 in 1970 for an electronic gas burner ignition system.

Electric stoves had to wait until the late 19th century, when Edison and his rivals began providing electricity to homes. These stoves were expensive, and their heat was poorly controlled. Prolific Canadian inventor Thomas Ahearn demonstrated an electric range at the Chicago World's Fair in 1893, but did not patent his design in the United States. The first American inventor to patent an electric stove was Charles Edward Roehl, in 1895 (patent number 532,909). Hundreds of other improvements and designs were generated in the next two decades, and starting in the 1920s, electric stoves become popular.

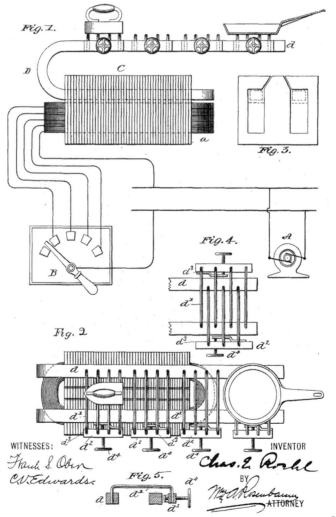

Patent no. 532,909

How Stoves Work

Stoves, traditionally heated by burning wood or coal, are now heated almost exclusively with either natural gas or electricity. Electric stoves generate heat by passing strong electric current through a highly resistive metal coil, or burner. Electric burners are made of a nickel-chromium alloy sheathed in stainless steel. The burner can heat up to become glowing or "red hot." (For more information on how heating elements work, see the introduction, p. xiii.)

The temperature in an electric stove is controlled via an "infinite switch" that regulates the flow of electricity to the heating coil. You rotate the dial on the switch to power the coil. An infinite switch works like a dimmer switch in that it controls the percentage of the alternating current cycle during which power is allowed to flow to the heating coils—the further you turn the dial, the more electricity flows.

Traditional gas stoves mix natural gas piped in from an outside source with oxygen from the air to burn. Older models require the gas stream to be lit by hand or by a pilot light. New models have built-in electronic starters that "click" on when you turn the dial to "light."

In addition to gas and electric ranges, other technologies are now available. Glass-ceramic stovetops easily pass infrared energy from electric heating coils or infrared lamps to the pots above. Other modern kitchens feature energy-efficient induction stoves; they generate heat by creating a rapidly oscillating magnetic field,

Glass-ceramic stovetop

which induces a corresponding magnetic field in an iron or steel pot placed on the range top. (A magnetic field cannot be induced in a copper or aluminum pot, so they will not heat up efficiently on an induction stovetop.)

Inside the Stove

Traditional electric stoves have high resistance heating coils that sit in openings in the stovetop. They are connected to electrical power through plugs, making it easy to remove them for cleaning or replacement. The electric coils are surrounded by ceramic material—you won't get shocked if you touch one. The infinite switch attached to the stove dial adjusts the flow of electricity to maintain consistent heat output.

Gas stoves may be lit via pilot light or electronic igniter; see the chapter on gas ovens (p. 115) for more information on how these components function.

Under the surface of a glass-ceramic stove, you can see either the

Heating coil and infinite switch

electric heating coils or the infrared lamp that generates the heat. The stove works like an ordinary electric stove, except that the heat generator is hidden out of sight, beneath the stovetop. Heat from the coils or lamp radiates to the pot above.

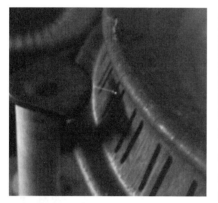

A spark jumps from an electronic igniter to the burner

Under the surface of an induction stove are coils of copper wire. High-frequency AC current is cycled through the coils to generate the strong magnetic field that produces the heat.

What's That Smell?

What do you smell when you turn on the gas range? It's not the natural gas you smell but mercaptan, an additive that doesn't contribute to combustion. It is added so you are aware that the gas, otherwise odorless, is on.

Teflon-Coated Frying Pan

History of Teflon

That nonstick Teflon coating on your frying pan is the result of one of the most celebrated mistakes in the history of invention. Roy Plunkett often told the story of how he discovered Teflon when he wasn't looking for it, didn't know what it was when he found it, and had no idea of what it could be used for. Plunkett, a research chemist working for a subsidiary of DuPont at its research lab in Deepwater, New Jersey, was researching new products to use as refrigerants when he made a curious discovery: a gas stored overnight in a pressurized cylinder had apparently disappeared. Plunkett didn't believe that it had escaped from the cylinder, and he confirmed his suspicion by weighing it. Cutting the cylinder open revealed a white powder that had strange properties—it didn't react with other chemicals, it had the lowest coefficient of friction of any tested substance, and acids and solvents didn't affect it.

Plunkett had followed the first law of science: when you find something interesting, stop what you're doing and check it out. He was awarded patent number 2,230,654 in 1941, and was later inducted into the National Inventors Hall of Fame for his discovery.

Manufacturers in the United States didn't believe that Teflon-coated pans would sell. Thomas Hardie, an American reporter for United Press International, discovered Teflon-coated pans while traveling in France. For years, he tried to interest U.S. manufacturers in making them, but no one was interested. So he purchased several thousand pans and imported them from Europe. Again, he had no luck getting stores to carry them—until 1960, when Macy's placed a small order. They sold out in a few days and the Teflon pan boom was launched.

Eventually it was discovered that Teflon decomposes at the high temperatures that may occur in cooking, and the chemicals released can be harmful. In addition, the Environmental Protection Agency announced that one of the chemicals used to manufacture Teflon is probably carcinogenic. So DuPont scientists have created other coatings (most notably Silverstone) that are now used in coating pots and pans.

How Teflon Works

Teflon pans keep foods from sticking by presenting a surface of inert molecules—molecules that don't bond to other substances. That raises the question: how do you get a substance that doesn't stick to anything to stick to the metal frying pan? This sticky issue was resolved by DuPont chemists: They first apply a very sticky primer to the metal pans and then add two or more layers of Teflon on top. In the manufacturing process the pan is heated, and this makes the primer bond to both the pan and the Teflon coatings.

Inside Teflon

Teflon is a long chain of molecules. The chain is composed of carbon atoms that are surrounded by fluorine atoms. The atoms of carbon and fluorine bond to each other very strongly. The strong bonds are what make Teflon so inert and slippery.

The Gecko Test

If a gecko can hang from your frying pan, it doesn't have a Teflon surface. Ask your local pet store owner if you can borrow a gecko to give this a test. Teflon is the only manufactured material that defies the sticky feet of the likeable lizard.

The gecko grips surfaces by employing millions and millions of tiny hairlike structures on its feet. These tiny hairs are attracted to the molecules of almost any substance the lizard stands on—any substance except Teflon.

TIMER

History of the Timer

Thomas Norman Hicks invented the mechanical kitchen timer in 1926 (patent number 1,602,260). Before that, cooks relied on hourglass-like devices—for instance, the "Improved Egg-Timer" developed by William Silver in 1895 (patent number 333,350). The electronics age has brought dozens of new designs to solve the problem of keeping track of boiling eggs.

How Timers Work

There are several ways to time cooking. The simplest is the gravity egg timer, more commonly known as the hourglass: sand or salt falls through a constriction in a glass tube. The rate of flow is controlled by the narrowness of the constriction; the total time is controlled by the amount of sand or salt inserted into the tube.

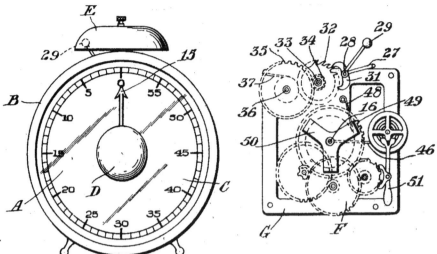

Patent no. 1,602,260

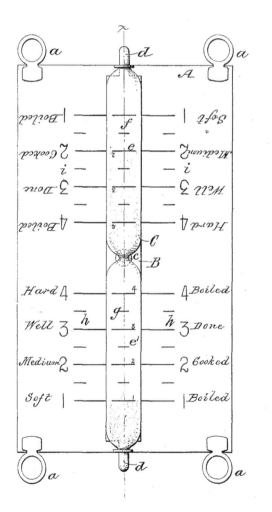

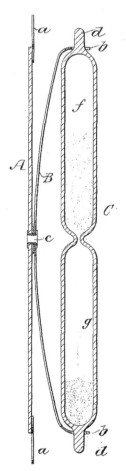

Patent no. 333,350

Mechanical timers are a bit more complicated. You provide the energy that runs the timer when you turn the dial against the resistance of a spring inside. The spring then unwinds, turning a series of gears to move the dial back toward zero and then ringing a bell to announce, "Time's up!" But without any other controls, this would provide erratic timekeeping. To control how fast the main spring unwinds, its gears are connected to a mechanism called an escapement. The escapement regulates the turning of gears by catching on a gear tooth and stopping the motion briefly. Then the escapement releases the gear, the main spring unwinds just a bit, and the escapement catches on the next tooth.

The timing of the escapement's catch-and-release cycle is regulated by either a pendulum or a balance wheel. A pendulum swings back and forth at a rate that depends on the length of the pendulum: how far its bob, or weight, is from the center of rotation determines how quickly the escapement catches and releases the gear. A balance wheel, on the other hand, spins in one direction and then in the other; its motion is regulated by a small spring called a balance or a hairspring. In the device I take apart below, a balance wheel and a small helical hairspring control the rate of movement of the escapement.

Electric timers that plug into household current may use small alternating current motors. The motors turn at a steady rate regulated by the frequency of the alternating current, which is closely monitored by power companies as they produce electricity. In the United States, AC power cycles 60 times a second, so the timer counts off one second for every 60 cycles of current.

Most modern timers are battery operated and use quartz timing. When exposed to the battery's electric current, a quartz crystal vibrates a consistent number of times per second. The timer's electronic circuit counts the number of vibrations and changes the LCD readout when the right number have occurred.

Inside the Timer

The dial pulls off, revealing a set of screws that allow the back to be removed. Three more screws allow the two inner halves to separate. I removed them and pulled the two halves apart. Once released from its confines, the spring jumps out.

The large spring stores the energy you input when setting the time. You turn past the time you need all the way to 60 minutes and then back to the desired time. This ensures that this main spring is wound tight. It unwinds through a series of gears that move the escape-

Timer interior with spring unraveled

ment back and forth. Each time the escapement stops moving, we hear "tick."

The rate of movement is controlled by the balance wheel. Beneath the balance wheel is a much smaller spring called a hairspring or balance spring. As the balance wheel spins, it winds the hairspring, which resists the motion of the balance wheel. So the balance wheel spins one way until slowed and reversed by the spring, and then it spins in the other direction. This motion would quickly stop if no additional energy were available, but that's what the escapement does—it moves power to the balance wheel.

The hairspring controls how quickly the balance wheel returns to the starting position.

Balance wheel and hairspring

Each time the wheel comes back, the escapement moves to spin it again. Tick. The hairspring regulates the timekeeping process. The main spring powers the motion. With motion at a constant rate, a gear rotates the dial. When the dial moves back to zero, a lever releases a metal-tipped plastic hammer that hits the metal housing: "Bing!" The eggs are cooked. (You positioned the hammer when you turned the dial to set the timer. If you turn the dial only a short distance, the hammer won't be locked into position and the timer won't sound when it winds down.)

Hammer (bottom) and metal housing (top)

TOASTER

History of the Toaster

Humans have been transforming bread into toast for centuries. The Romans learned about toasting bread from the Egyptians, and they carried this knowledge with them as their empire spread throughout Europe.

Until recently in bread history, the person making toast held a piece of bread in front of the coals of a fire. They supported the bread, and avoided burning the hair on their hands, by using first wood and later metal sticks.

Even before Albert Marsh invented Nichrome heating wire in 1905 (see the introduction, p. xiii), people used rudimentary forms of heating wires to make toasters. In 1895, Levi Edwards got a patent for his toaster/broiler (patent number 533,795). But until Marsh's innovation, the technology was unreliable, and heating elements burned out quickly. George Schneider, working only a few blocks away from Marsh, took advantage of Nichrome's durability by incorporating it into several different products. One was an "electric cooker," patented in 1906 (patent number 825,938). His patent does not mentioning toasting, but some historians credit this as the first electric toaster.

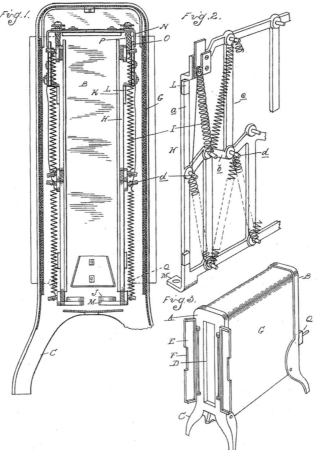

Patent no. 825,938

General Electric claims to have made the first electric toaster in 1905, but their earliest toaster patents weren't issued until several years later, when Harold Bradley and Frank Shailor received patents and assigned them to GE in 1909 and 1910, respectively (patent numbers 926,714 and 950,058). Albert Marsh had patented his own toaster in 1907 (patent number 852,338), but he ultimately decided to concentrate on making his heating wire for other manufacturers to incorporate into their products.

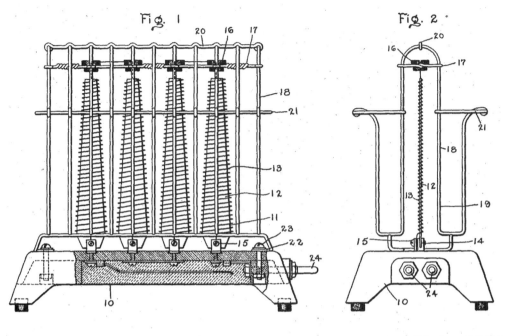

Patent no. 950,058

Initially, electric toasters were found only in restaurants, as most homes didn't have electricity. As electrification swept the country a few years later, these inventions finally caught on with consumers. Early models required the cooker to turn the bread slice around and to remove it before it got burned. Later models allowed the user to toast both sides simultaneously, and incorporated timers and thermostats.

Then, in 1921, Charles P. Strite invented the pop-up toaster (patent number 1,394,450). His invention included a timer and a spring to launch the toast upward when the specified time had elapsed. From a timer to a heat sensor was the next big step. The first thermostat-equipped toasters sensed the temperature of the heating element; Sunbeam improved the thermostat so it sensed the temperature of the bread.

Today, toasters are high-tech devices, built to incorporate heat-resistant plastics and solid-state electronics.

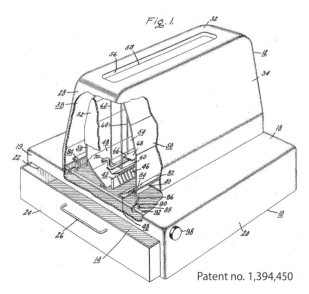

Patent no. 1,394,450

How Toasters Work

Toasting a piece of bread doesn't just make it darker—it fundamentally changes the bread's composition. At 310° F, the toaster causes a chemical reaction called a Maillard reaction, by which amino acids and sugars in the bread react with each other. This is what produces toast's sweeter taste and great smells.

Electric toasters generate the necessary heat by passing electric current through high resistance Nichrome wires. Although there are lots of different styles, the most common is the pop-up toaster. (See the next chapter, p. 171, for a discussion of the toaster oven.)

> To keep your toaster working well, clean out the toast crumbs from breakfasts past. Accumulating crumbs interfere with the machinery.

After dropping a slice or two of bread or a couple of frozen waffles into a pop-up toaster's slots, you depress the lever. Often, metal grates press in against each slice to center it in the heating compartment. Pushing the lever all the way down engages a latch (either electric or mechanical) that holds it down against the upward force of a spring and starts the electric current flowing through the heating coils. In some models it also starts a timer; after a certain amount of time, the timer will shut off the current flow to the coils and release the latch so the toast can pop up. However, today more models use a thermostat instead of a timer; it disengages the latch and shuts off the current when the bread reaches a certain temperature. The user can adjust the toaster's time or temperature setting.

A typical toaster uses 600 to 1,200 watts of electrical power to perform its magic.

Inside the Toaster

A few screws held the plastic body onto the metal workings of this toaster. The switch at the back of the toaster that provides power is a beautifully simple lever that presses two copper fingers onto corresponding contacts. The lever arm is a metal rod that runs beneath the heating compartments of the toaster to another lever arm on the front. Pushing the knob down to start the process depresses the front lever arm, which rotates the rear arm, pushing the electrical contacts together. Electricity is now flowing to the heating elements. Each heating element consists of a single

strand of high resistance wire that is wrapped around a metal plate.

In this cheap toaster the spring-loaded knob assembly is held in the depressed position by a latch that slides into an indentation in a metal rod. As you depress the knob, the latch pushes against the rod until it falls into the indentation. The knob and toast can't rise until the thermostat, a simple bimetallic device that bends when heated, pushes the latch out of the notch. A coil spring launches the knob assembly and the toast back up.

Metal plate and resistance wire

Many other toasters hold the knob assembly down with an electromagnet instead of a mechanical latch. As long as electricity is flowing to the heating wires, it also flows to the electromagnet. When the thermostat heats up and bends, it cuts power to both components, causing the electromagnet to weaken and allowing a spring to launch the knob assembly and the toast skyward. Some electromagnetic toasters use an electronic circuit to time the toasting cycle. You adjust the timer by rotating a variable resistor, a device that changes how long the attached capacitor takes to charge. Once the capacitor is fully charged, the circuit cuts power to the electromagnet.

A friend donated a four-slot toaster he had used for 50 years. It still worked, which was one of several things that made it remarkably different from modern toasters. Other striking differences included the weight: it was much heavier than a comparable toaster sold today. The power cord was wrapped in cloth, not rubber, and the screws holding it together (there were 14 of them) were all flathead screws. Henry Phillips had invented his now-ubiquitous crosshead screw only a few years before this toaster was designed, so the older flathead screw was still preferred.

Knob assembly

Vintage toaster

Toaster Oven

History of the Toaster Oven

The first patent for a combination toaster and cooker was awarded to Adolph Osrow in 1938 (patent number 2,121,444); his patent describes the machine as being able to toast bread "while coffee is being made or chops, cereal or other articles cooked upon the [built-in] grill." Two years earlier, Calkins Appliance Co. began selling their own toaster oven, known as the Breakfaster, but it wasn't patented until 1946, when Otto Stelzer was awarded a design patent that protected the look of the device rather than its function (patent number D145,108). Nevertheless, judging by the number of vintage Breakfasters being offered for sale today on the Internet, this must have been a successful line of products at the time. (For photos of the Breakfaster, see the Web site for the Cyber Toaster Museum, www.toaster.org/1920.php?page=5.)

The machine that sparked popular interest in toaster ovens was submitted to the Patent Office in 1954 by Wilbur Schmall. It had two different chambers, one for toasting

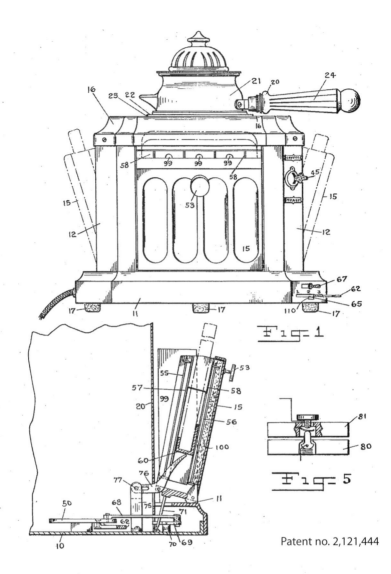

Patent no. 2,121,444

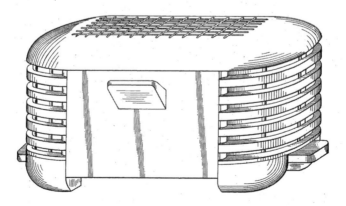

Patent no. D145,108

and one for baking; the user selected which one he or she wanted to use. Schmall assigned the patent to General Electric, which incorporated modifications from other inventors and used the revised design as the basis for a new product line: the T-93 toaster oven, released in 1956. Two years later, the Patent Office officially registered the patent for Schmall's original design (patent number 2,862,441).

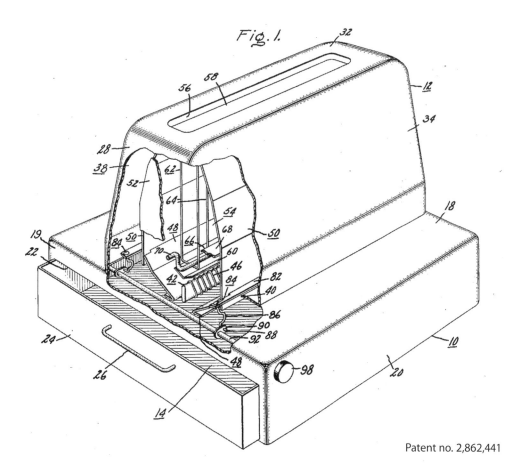

Patent no. 2,862,441

How Toaster Ovens Work

Unlike most toasters, into which you drop one or more slices of bread vertically, toaster ovens toast on the horizontal plane. Bread to be toasted or leftovers to be heated sit on a grid or tray—and thankfully neither one pops up when the food is done! When you select "bake," "broil," or "toast," heating coils above and/or below the bread come on. (For more information on how heating elements work, see the introduction, p. xiii.) For baking, only the bottom heating element is energized; for broiling, only the top one is; toasting uses both heating elements so both sides of the bread can be toasted.

An electronic or mechanical thermostat controls the current in the heating coils. A toaster oven uses more power than a traditional toaster.

Inside the Toaster Oven

Removing a few screws allowed the outer metal jacket to come off. The cooking compartment itself contains the heating rods and the trays, but most of the other interesting stuff is mounted on the inside of the door frame.

At the top is the dial to select the function—"bake," "broil," or "toast." As you turn it, it rotates a rod mounted with three irregularly shaped pieces, or cams. Only two of them are functional. When rotated into the proper position, the two cams push electrical contacts together. When the toaster is set to "off," neither cam is pushing against a contact. In "bake" mode, the inner cam pushes the inner contact switch closed. This allows current to flow to the bottom heating rod. For "toast" mode, the second cam also closes the outer switch, so both heating elements can engage. Turning to "broil" keeps the outer cam on the outer switch, but rotates the inner cam so its switch is open, allowing only the upper heating rod to activate.

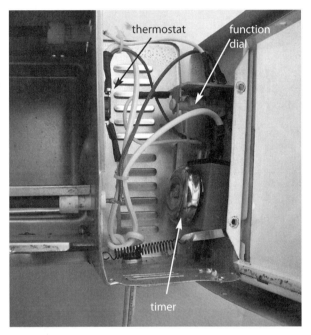

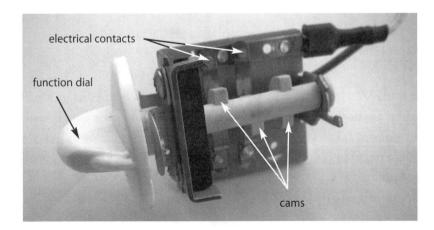

Also attached to the door frame is a mechanical timer; you twist the dial to start electricity flowing to whichever heating rod is engaged. When the timer has run down, power is cut and the bell sounds, telling you that the food is ready.

Opposite the other controls, mounted on the outer wall of the cooking compartment, is the thermostat that maintains the heat in the cooking chamber at 310° F, the temperature at which bread is toasted. Whenever the temperature of the thermostat rises above 302° F, it shuts off power

Timer

to the heating rods, and it powers them on again when the temperature drops. Why 302° F and not 310° F? The design engineers took into account the fact that the thermostat will be a few degrees cooler than the toast in the cooking compartment.

Thermostat

TRASH COMPACTOR

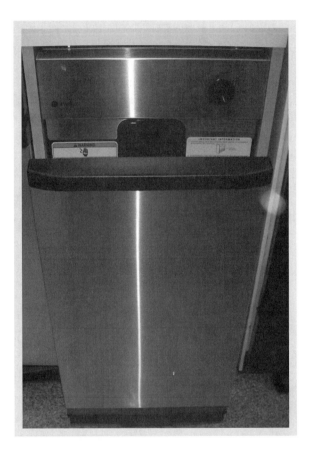

History of the Trash Compactor

Inventor M. S. Wells patented a hand-operated "can crusher and baler" in 1941 (patent number 2,234,098). Sixteen years later, two inventors from South Bend, Indiana, patented the motor-assisted "can and bottle crushing and disposal machine" (patent number 2,800,159). Their patent application suggested that their invention "serves as a reliable and practical means to appreciably assist one in coping with and systemically handling the every-day problem of expeditiously disposing of used commodity and beverage containers," which they went on to list.

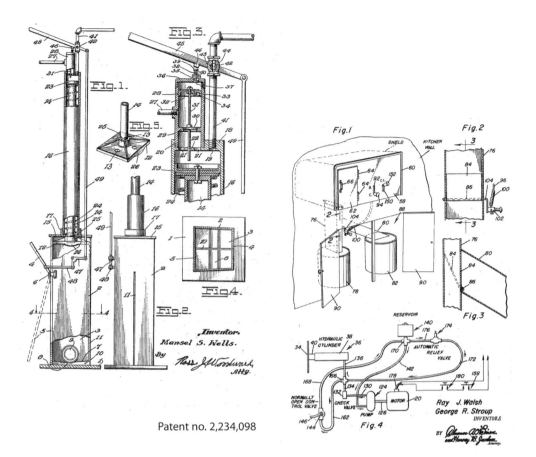

Patent no. 2,234,098

Patent no. 2,800,159

Compactors have been slow to catch on. Though once favored for their ability to save space in garbage dumps, many in the recycling age consider them self-defeating appliances. Only about 3 percent of American homes have them.

How Trash Compactors Work

A trash compactor can compress five bags of garbage into one bag. To operate, you drop trash into a disposable bag that sits in a metal bin inside the compactor. When you're ready to squish it, you close the compactor and push a switch. The switch activates a motor in the base or the top of the compactor.

Base-mounted motors are connected to a chain that drives two vertical screws. The screws, in turn, support a steel "ram" that hangs over the metal bin. As the screws rotate, the ram is pulled down, compressing the cottage cheese containers and chicken bones into a crumpled mess sure to frustrate future archaeologists.

Top-mounted motors turn a metal screw above the ram, which presses rather than pulls the ram downward. In this design the ram cannot travel downward very far due to the short length of the screw. So it can only compact the trash when the bin is nearly full.

When the ram has passed its "squished" point, it flips a mechanical switch that reverses the motor. The screw or screws rotate in the opposite direction, raising the ram.

Inside the Trash Compactor

With the door and metal bin out of the way, you can see the ram, the switch that reverses the direction of the ram, and an aerosol spray can of air freshener. Each time the compactor runs, it pushes the nozzle of the can, providing a shot of air freshener to mask the odor of the garbage being compacted.

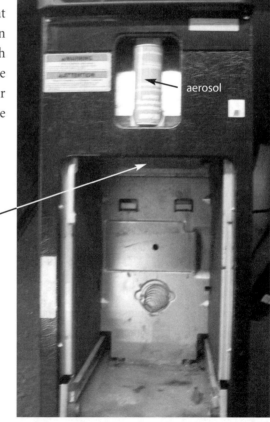

TURKEY TIMER

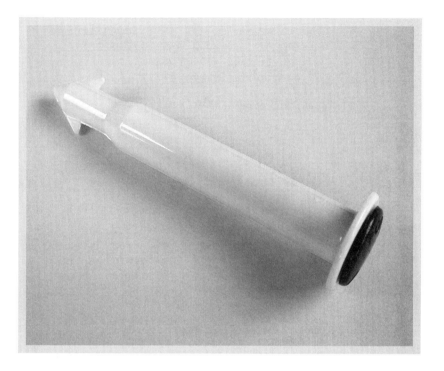

History of the Turkey Timer

In the 1960s the California Turkey Producers Advisory Board had a problem. People complained that their cooked turkeys were too dry. The board figured out that many people were cooking their turkeys too long, so they brainstormed how to create a device that would indicate when a turkey was ready to be taken out of the oven. One member of the board, George G. Kliewer, glanced up at the fire sprinklers in the ceiling and realized that they were triggered when heat from a fire melted a piece of metal inside the sprinkler. The board realized that the same principle could be used to determine when a turkey reached a particular temperature. Under the leadership of another board member, Eugene Beals, they adapted the concept and started testing possible designs, a process that continued for a year.

Once they had a working design, they formed a company, Dunn-Rite, which was later sold to 3M and then to Volk Enterprises. George Kliewer shows up as the sole inventor on the 1966 patent (patent number 3,280,629). The timers sold today have been improved several times from this early design—they use different combinations of materials to achieve more precise temperature activation.

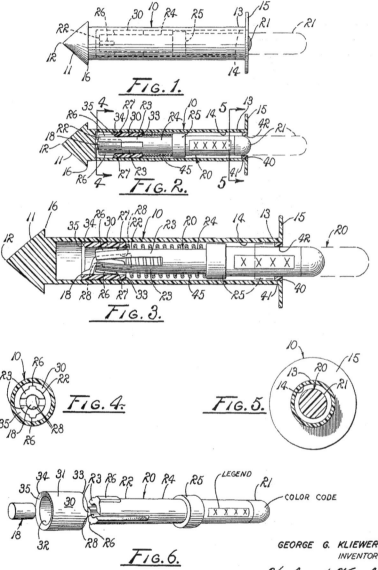

GEORGE G. KLIEWER
INVENTOR

Huebner & Worrel
ATTORNEYS

Patent no. 3,280,629

How Turkey Timers Work

The turkey timer is a spring device made of plastic that is inserted into the turkey before it's cooked. The pop-out indicator in the center is held in place by a small dab of soft metal. When the turkey is finished cooking—when the inside temperature reaches 185° F—the metal melts, freeing the indicator to move. A compressed spring pushes the now-free stick into the "cooked" position.

The patent explains that the "fusible latch element," the metal that melts at the predetermined temperature, can be made of a metal alloy of bismuth, lead, tin, and indium or cadmium. (These materials are encased in a plastic housing to keep them from touching the turkey.) By varying the percentages of each material in the latch, the indicator can be configured to release at different temperatures, from 136° to 203° F.

Inside the Turkey Timer

To reverse engineer a turkey timer, I held the pointed end of the timer in a pot of boiling water. Within 15 seconds, the indicator popped out. Then I cut open the side of the timer to release the indicator and reveal the spring.

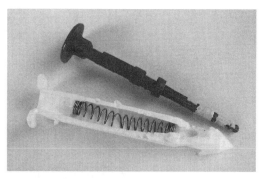

At the bottom end of the indicator are two bands of the material that melts at the predetermined temperature to release the indicator. This disposable timer was made by the Heuck Company; it probably uses a different set of materials for the fusible latch than what's described in the original patent.

Don't Throw It Out

What do you do with a used turkey timer? Reuse it! Submerge the pointed end of the timer in boiling water while pushing down on the popped-out top of the indicator. After a few seconds, the metal in the base of the timer will melt. Remove the timer from the water and let it cool while continuing to hold the indicator down against the force of the spring. In a few seconds the metal will resolidify and you can release the indicator. You're now ready for next Thanksgiving.

WAFFLE IRON

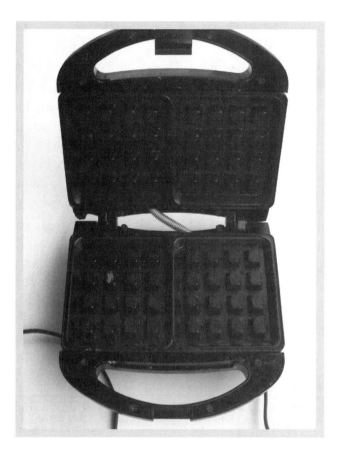

History of the Waffle Iron

Waffle irons were first used in Europe in the 1300s. The cook filled a special plate with batter, hooked it together with another plate, and held the assembly over a fire with two long wooden handles. When the Pilgrims came to the New World on the Mayflower, they brought their waffle irons with them.

The first waffle iron developed in the United States was patented by Cornelius Swartwout in 1869 (patent number 94,043). It was designed to sit on a wood-burning stove. In the early 20th century, the thermostat-controlled heating element was invented, allowing the creation of electric waffle irons. William D. Wright obtained the first patent for an electric waffle iron in 1917 (patent number 1,214,486), but by 1911 General Electric was already making electric waffle irons, possibly protected by an earlier patent that would be difficult to locate because it made no mention of "waffles."

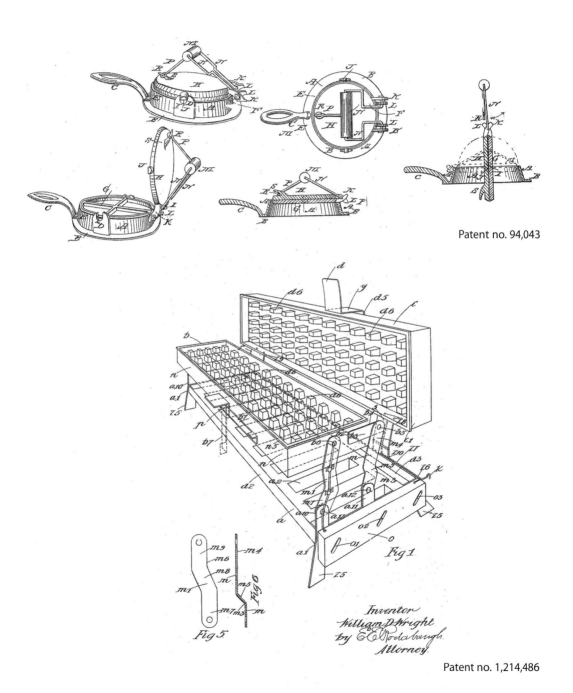

Patent no. 94,043

Patent no. 1,214,486

Frozen waffles, which don't require the services of a waffle iron, were introduced in 1953. The inventor, Frank Dorsa, also invented an automatic, continuous potato peeler to use in the manufacture of potato chips.

How Waffle Irons Work

The waffle iron's design makes it easier to cook batter to a crispiness that begs for maple syrup. The indentations in the grill plate help better distribute heat throughout the cooking batter.

You preheat the waffle iron before pouring in the batter. In some models, an indicator light goes out when a thermostat senses that the waffle iron is hot enough. It comes on again when you pour in the batter, and goes out a second time when the waffle is cooked.

Heating the batter drives out moisture, which you can see as steam escaping from the waffle iron. Heat also rearranges the proteins in the flour and eggs. These proteins form solid bonds and hold the waffle in its trademark gridlike pattern.

Inside the Waffle Iron

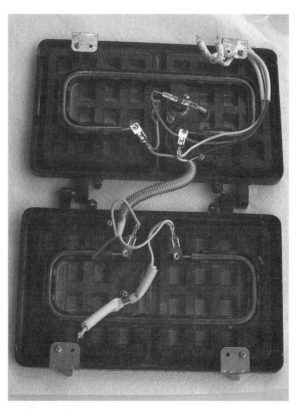

A few Phillips screws held the plastic shell, top and bottom, to the metal working components. The upper and lower griddles are almost identical. Nearly oval-shaped tracks in each griddle hold the heating elements; they are firmly anchored and defied my attempts to set them free. (For more information on how heating elements work, see the introduction, p. xiii.) Electrical power is supplied to each side, but the upper one has a thermostat, which controls the power flowing to the two indicator lights.

The thermostat is a bimetallic thermal switch. It is held against the underside of the upper griddle so it will heat up along with the griddle. As it does, the metal inside the thermo-

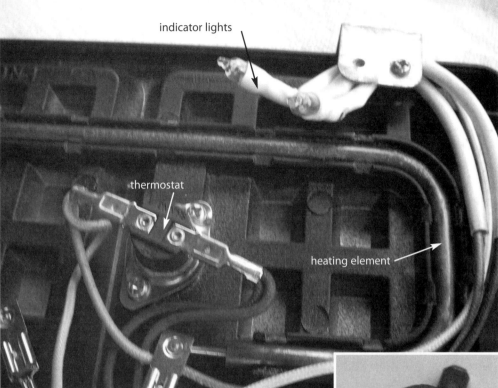

indicator lights

thermostat

heating element

stat expands and interrupts the electric current both to the heating elements and to the indicator lights that shine through the translucent cover in the lid, indicating that the waffle iron is hot enough. When you add the waffle batter, it reduces the temperature of the waffle iron, and the thermostat restores power to the lights and heaters. When the waffle is cooked to the proper temperature, the thermostat again switches off.

Thermostat

These Waffles Taste Like Rubber

Bill Bowerman found a novel use for waffle irons in 1971. The University of Oregon track coach poured latex rubber into a waffle iron to make a new sole for running shoes. When his experiments proved successful, he and his business partner, former track runner Phil Knight, produced and sold the shoes through their company, Nike.

WATER FILTER

History of the Water Filter

Water filters are not new, but modern home filters that fit on kitchen faucets are. The earliest patent for a water filter was awarded in 1845, long before homes had faucets, to Joseph Craddock (patent number 4,344). His invention was both a water filter and a water-cooler—see the watercooler chapter (p. 193) for the illustration. Craddock's invention used gravel and coal as the filtration media.

The first faucet-mounted filter patent was award in 1892 to Luther Blessing (patent number 471,840), but it wasn't until 1979 that Thomas Corder patented the faucet-mounted filter cartridge and base (patent number 4,172,796) that appears to be the basis for the filters in use today.

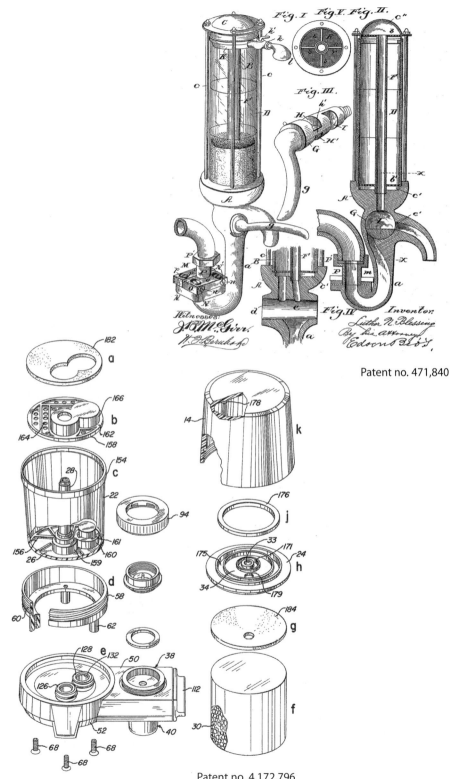

Patent no. 471,840

Patent no. 4,172,796

How Water Filters Work

Even if the water leaving your local water plant is pure, it can pick up a variety of chemicals on the way to your faucet that you would be better off not consuming. A faucet filter removes them.

There are different ways that a filter can purify water. None are perfect, but each removes some of the potential impurities. If you are considering purchasing a filter, first find out what your water quality is so that you know what impurities you need to remove. If you are using water from a municipal water supply, they can give you a report on water quality. As an additional check, you can send a water sample from your kitchen tap to a company that analyzes it and gives you a report.

The most common type of water filter is the activated carbon filter. This filter consists of a bed of finely ground charcoal that has been "activated" by heating it with various gases, which opens up numerous tiny pores in its surfaces. As water passes through the filter, the pores capture particles, chemicals, and organic compounds. Key to the success of the carbon filter is how finely the charcoal is ground; a finer grind creates a larger surface area with which to catch the impurities.

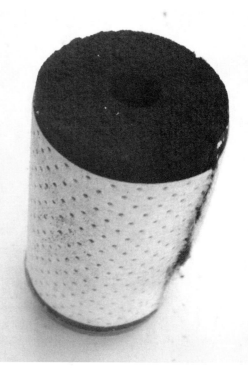

Activated carbon filter

Carbon filters are less successful at removing bacteria and metals like lead, unless other materials are added. For example, Corder's patent discusses the use of "oligodynamic silver" in the filter medium to kill bacteria—ions of heavy metals can destroy proteins in living cells via what's known as the oligodynamic effect. To remove lead, zeolites are added to the activated carbon. These are molecular sieves—substances with tiny openings that can filter out molecules based on their size or shape.

A different but still common kind of water treatment is an ion exchange filter. It makes water softer by removing calcium and magnesium. If your soap doesn't lather easily, your water is probably hard and an ion filter may help. More complicated are reverse osmosis filters, which are very effective at removing most—but not all—contaminants. Distillation filters kill bacteria and remove most organics and metals, but not all pesticides or volatile chemicals. Ceramic water filters use ceramic materials containing millions of very small holes through which the water is forced to pass. Materials larger than a water molecule are stopped by the filter. The ceramic filter must be cleaned periodically to flush away all the filtered materials.

Inside the Water Filter

The faucet-mounted water filter has two components: the replaceable filter cartridge and the base that attaches to the faucet. In this model, the filter cartridge sits on top of the

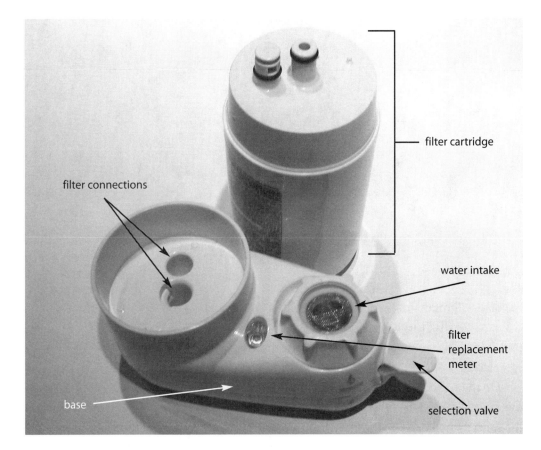

base. Water from the tap flows into an intake opening, through the base, and up into the filter cartridge. There it passes through the filter medium, travels back into the base, and flows through a second opening into your pitcher.

In the center of the base is a "filter replacement meter." It tracks how much water has passed through the disposable cartridge and alerts you when you need to replace it. This model measures out 100 gallons before changing the color of the indicator from green to red, indicating that it's time to change cartridges. How does the indicator work? As water flows through the filter, it spins a turbine in the center of the base. The turbine is geared to greatly reduce the speed of rotation. Connected to the gearing is a disk that is colored green, except for one section that is colored red. The indicator disk slowly turns as the water passing through the system spins the turbine. Eventually, the red section of

Meter turbine (center)

Indicator disk and spring

the disk will rotate into place under the indicator's clear cover. As the disk turns, it also winds up a spring that will pull it back to its starting position when the cartridge is replaced.

On the end of the base is the valve that lets you select filtered water or tap water. When the valve is turned to the tap water option, water flows from the tap straight down into the sink. Turned to filtered water, the valve sends water through the turbine and into the filter. The black bands on the valve stem are O-rings. They form a seal between the stem and the valve body.

Inside the cartridge is a cylinder of activated charcoal wrapped in a paper sleeve with holes. Water comes into the cartridge along the outer edge and passes between the cartridge's outer casing and the sleeve. It flows through the holes in the sleeve and through the charcoal. Water pressure forces it to flow toward the opening in the center of the charcoal cylinder and out the bottom of the cartridge.

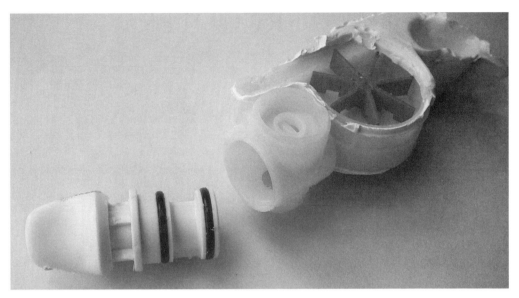

Selection valve (left)

Filter cartridge

Build Your Own

The water filter is one of the few kitchen appliances that you can make yourself. I'm not recommending that you do this—but you could. The easiest water filter to make is a drip flow filter, which water falls through under the force of gravity. On the way down, the water passes through one or more filtration beds. Sand and charcoal make good filter media; they can remove larger particles and many chemicals impurities. (If this sounds quaint, consider that many municipalities use sand filters to treat their water supplies.) The filtration medium is contained within a bag of fine cloth, which keeps it from being carried along with the water.

Place the cloth bag within a large, waterproof container—a large bucket, for example—with a hole in the bottom. Fill the bucket with water and let it flow through the bag and filtration medium and out the hole. It takes a couple of weeks to "season" the filter medium with bacteria that will remove organic material (the bacteria will take up residence in the filter as long as it is exposed to air), but a sand filter will work right away to clean chemical impurities from the water.

WATERCOOLER

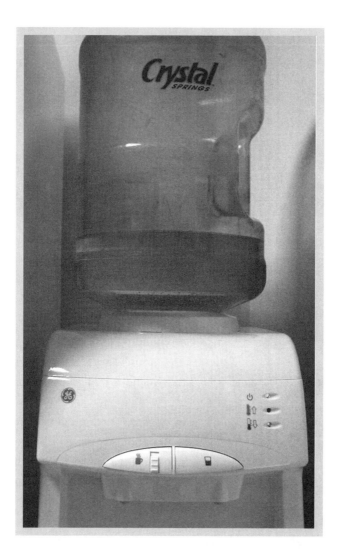

History of the Watercooler

Some of the earliest inventions in the United States were watercoolers and water-cooled refrigerators. In 1845, Joseph Craddock was issued patent number 4,344 for a device that both filtered water and cooled it with ice. In 1857, James Harrison developed a mechanical refrigeration system based on vapor compression; the same technique was later applied to the cooling of water. Edward Williams (patent number 2,009,949) invented a vapor-compression watercooler in 1935, and Ralph Billings of General Electric patented an improved model in 1939 (patent number 2,160,639).

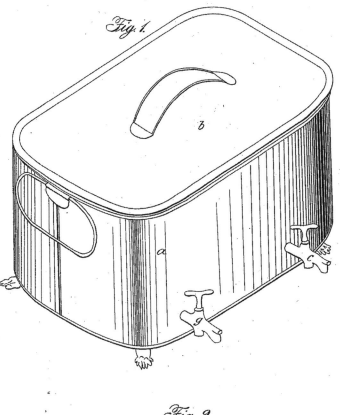

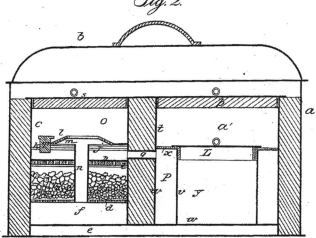

Patent no. 4,344

However, vapor-compression coolers tended to be complex and bulky. Smaller home units were made possible in 1963, when Max Alex was awarded patent number 3,088,289 for developing a watercooler that used a simpler, more compact refrigeration system based on the thermoelectric cooling effect.

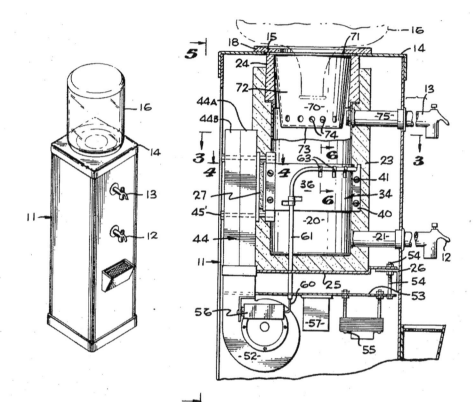

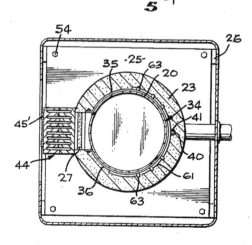

MAX ALEX
INVENTOR.

BY

Mason & Graham

Patent no. 3,088,289

How Watercoolers Work

Once a week or so, a delivery truck drops off a 40-pound, five-gallon container of spring water at your home or office. You place the jug upside down on the cooler stand. Older models require you to first open the jug, then flip the 40-pound container upside down and position it in the cooler quickly so you don't spill (much) water. Newer models don't require uncapping the jugs beforehand. Instead, the seal is punctured as you lower the bottle onto the cooler, which prevents spills. Once the jug is in position, gravity feeds the water through a valve that lets out water and lets in air (those big bubbles you see rising in the jug when you fill your cup) so that the jug doesn't form a vacuum, which would prevent any more water from coming out. To get water from the cooler, you raise a lever to open the tap and out it comes. Most coolers have two taps, one that provides cooled water and the other that provides either room-temperature or heated water.

Some of today's watercoolers still use vapor-compression technology (see the chapter on refrigerators, p. 131, for more information), but more and more small coolers chill the water by means of a thermoelectric cooling system. This system uses solid-state materials and an electric current to transfer heat from one place to another. The system works on the Peltier effect, which occurs when electrons move between two materials of differing electron density. When electrons are forced to move from a region of higher density to a region of lower density, they jump to a higher energy state by absorbing thermal energy, thus cooling this area. When they're forced from a region of lower density to a region of higher density, they drop to a lower energy state and release their heat. In a thermoelectric cooling system, the electric current drives electrons from one material to the other, heating one and cooling the other. The cooling side is placed adjacent to the water in order to chill it, and the warming side is allowed to cool through contact with air from the surrounding room.

The electronic controls regulate the cooling cycle based on the temperature in the insulated cooling chamber. A temperature sensor called a thermocouple monitors the temperature.

Inside the Watercooler

The small cooler I disassembled contains a thermoelectric cooling system that cools only a small amount of water at a time. The body of the cooler is held together with a few screws. Removing them allows the two halves to come apart. The receptacle at the top holds the water jug. Water enters the receptacle and flows either to the room-temperature tap or to the cooling mechanism just below. On the mechanism's outer surface, the cooling fan and cooling ridges are visible. They remove heat from the hot side of the cooling mechanism.

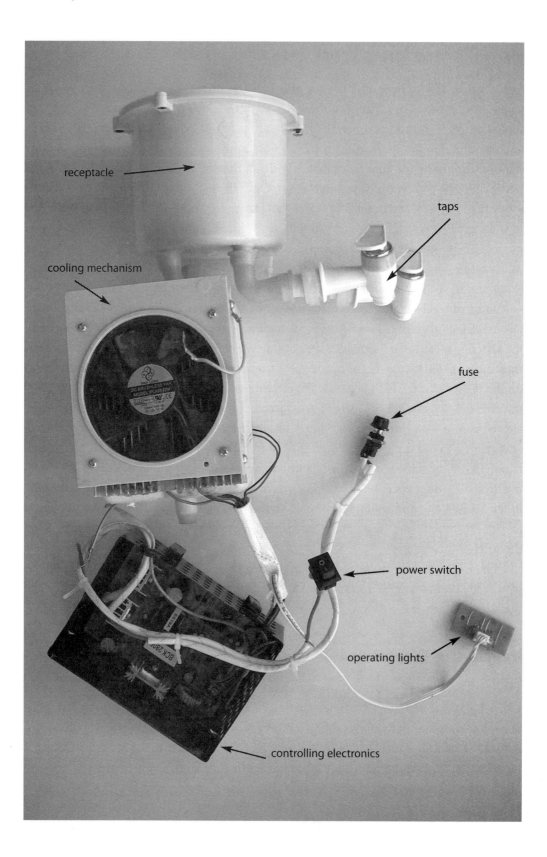

receptacle

taps

cooling mechanism

fuse

power switch

operating lights

controlling electronics

Removing the cooling ridges and fan reveals the insulated chamber where water is cooled and stored. The cooling element sits on a metal insert that fits into the chamber. The thermocouple fits into the opposite side.

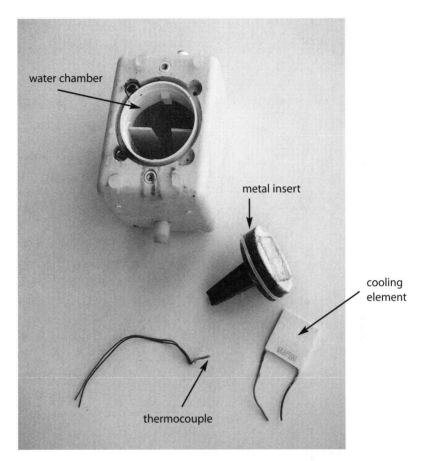

The cooling mechanism is connected to the cooler's controlling electronics, as are the power switch, fuse, and operating lights (LEDs). The fuse is probably to prevent further damage should the circuit get wet.

A Source of Electrical Energy

The thermoelectric cooling effect can be reversed: instead of using electricity to generate a thermal difference, thermal differences can be used to generate electricity. Some engineers are looking at this as a possible way to take advantage of naturally occurring temperature differentials to provide power.

WINE SAVER

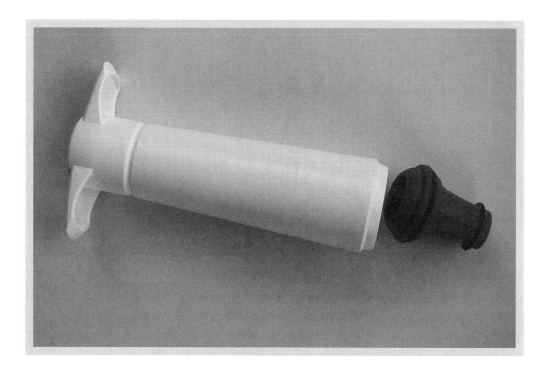

History of the Wine Saver

The idea of using a stopper and a pump to remove air from a wine bottle was patented by Bernardus Schneider in 1988 (patent number 4,763,803).

How Wine Savers Work

Once a bottle of wine is opened, it begins to oxidize. Oxygen in the air combines with some of the components of the wine, and this leads to a loss of color and flavor. To prevent this, an open wine bottle should be stopped up with a cork, and as much air as possible should be removed from the bottle. A pump and a corresponding valve/stopper, collectively known as a wine saver, can do the job.

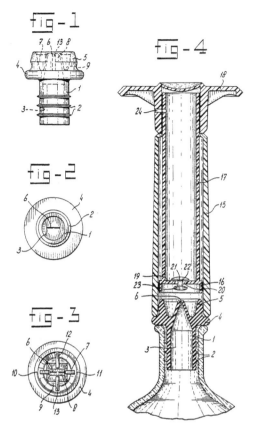

Patent no. 4,763,803

The stopper may be constructed according to several different designs. The valve may consist of a round seal or two rubber leaves that flex up and down. In either case, the valve opens when you attach the pump and pull on the handle, allowing you to draw air out of the bottle. When you stop pulling, the partial vacuum in the bottle draws the valve closed again, which prevents the air from returning to the bottle.

The pump is a simple affair, with a check valve that prevents air from escaping as you pull the handle and allows it to escape as you press the handle down again. On some models, the check valve clicks when you have extracted as much air from the bottle as you can.

You can test a pump's effectiveness by putting the end of it on your palm and pulling up on the handle. You should be able to feel the suction on your hand.

Inside the Wine Saver

This model's stopper is topped with a flexible rubber seal that allows air to leave the bottle when the pump handle is pulled. When the pump is fully extended, the lower pressure in the bottle pulls the flexible seal down to cover the four holes in the hard plastic cap beneath it, closing the stopper valve.

Rubber seal (right) and plastic cap (left)

The pump is made of ABS plastic and polypropylene. Sawing through the shaft holding the handle allowed the attached sliding cylinder to fall out of the outer tube. A rubber O-ring around the end of the inner cylinder prevents air from leaking out between the two tubes.

While the pump handle is being pulled out, the check valve at the very end of the inner cylinder seals to prevent air from escaping. As the handle is pushed in again, the check valve allows the air that was evacuated from the bottle to escape into the inner cylinder of the pump. From there it passes through two small holes on the inner cylinder, into the space between the two cylinders, and out of the pump.

When the pressure inside the bottle has equalized with the pump's suction pressure, a small piece of metal in the check valve will bend back and forth on each additional stroke, producing a clicking noise that lets you know you can't remove any more air.

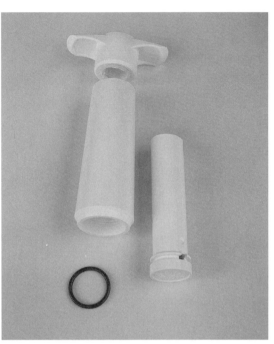

Disassembled pump

Check valve

BIBLIOGRAPHY

Ehrlich, Robert. *Turning the World Inside Out and 174 Other Simple Physics Demonstrations*. Princeton, NJ: Princeton University Press, 1990.

Inventive Genius. Alexandria, VA: Time-Life Books, 1988.

The National Inventors Hall of Fame Black Book. 27th ed. Akron, OH: National Inventors Hall of Fame, 1999.

Panati, Charles. *Extraordinary Origins of Everyday Things*. New York: Harper Perennial, 1987.

Sobey, Ed. *A Field Guide to Household Technology*. Chicago: Chicago Review Press, 2007.

Sobey, Ed. *A Field Guide to Office Technology*. Chicago: Chicago Review Press, 2007.

Walker, Jearl. *The Flying Circus of Physics*. New York: Wiley, 1977.

The Way Toys Work

The Science Behind the Magic 8 Ball, Etch A Sketch, Boomerang, and More

Ed Sobey and Woody Sobey

A Selection of the Scientific American Book Club

"This book is sure to provide hours of entertainment and enlightenment."

—*School Library Journal*

978-1-55652-745-6
$14.95 (CAN $16.95)

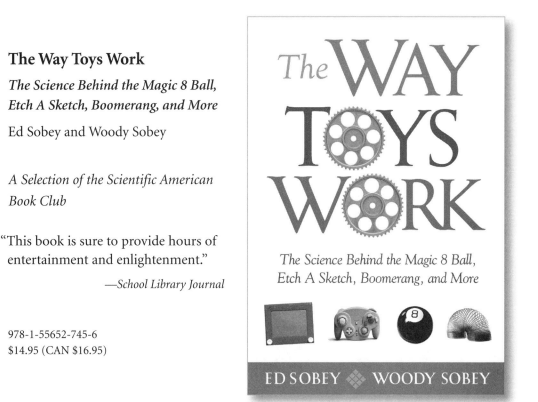

How does an Etch A Sketch write on its gray screen; why does a boomerang return after it is thrown; and how does an R/C car respond to a radio-control device? Father/son author duo Ed and Woody Sobey explain the science hidden in these and dozens more of the world's most popular toys. Each of the 50 entries includes the toy's history, patent application drawings, interesting trivia, and a discussion of the technology involved. The authors even include pointers on how to build your own using recycled materials and a little ingenuity, and tips on reverse engineering old toys to get a better look at their interior mechanics. Readers will also enjoy photos of the "guts" of the devices, including a filleted Big Mouth Billy Bass. The only thing you won't learn is how the Magic 8 Ball is able to predict the future—some things are better left a mystery.

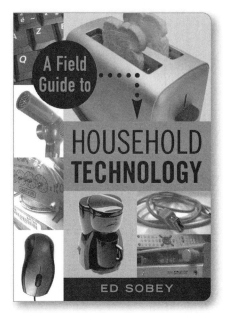

978-1-55652-670-1
$14.95 (CAN $18.95)

A Field Guide to Household Technology

Ed Sobey

Illustrating how a bathroom scale measures body weight and what the "plasma" is in a plasma-screen television, this fascinating handbook explains how everyday household devices operate. More than 180 different devices are covered, including gadgets unique to apartment buildings and houseboats. Devices are grouped according to their "habitats"— living room, family room, den, bedroom, kitchen, bathroom, and basement—with detailed descriptions of what they do and how they work, and photographs for easy identification. You'll never look at that pile of remote controls on your coffee table the same way again.

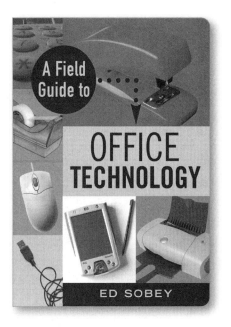

978-1-55652-696-1
$14.95 (CAN $18.95)

A Field Guide to Office Technology

Ed Sobey

The modern office can be a confusing technological landscape. How does a motion detector spot intruders? What is a voltage surge, and why does a computer need to be protected from it? And why do telephone keypads have their 1s in their *upper* left corners, while calculator keypads have their 1s in their *lower* left corners? Entries for more than 160 devices tell you how they work, who invented them, and how their designs have changed over the years. No longer will you need the IT staff to explain that mysterious blinking box in the coat closet.

A Field Guide to Roadside Technology

Ed Sobey

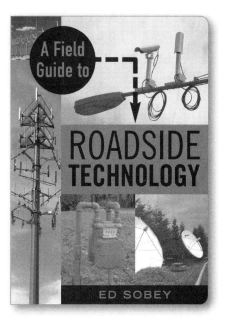

"Fun, informative, and easy to use."

—School Library Journal

If you've surveyed the modern landscape, you've no doubt wondered what all those towers, utility poles, antennas, and other strange, unnatural devices actually do. In *A Field Guide to Roadside Technology*, more than 150 devices are grouped according to their "habitats"—along highways and roads, near airports, on utility towers, and more—and each includes a clear photo to make recognition easy. Once the "species" is identified, the entry will tell you its "behavior"—what it does—and how it works, in detail.

978-1-55652-609-1
$14.95 (CAN $20.95)

A Field Guide to Automotive Technology

Ed Sobey

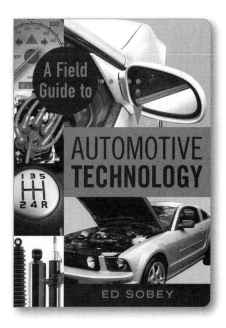

If you don't know your catalytic converter from your universal joint, *A Field Guide to Automotive Technology* is for you. How does an airbag know when to deploy? What is rack and pinion steering? And where exactly does a dipstick dip? More than 120 mechanical devices are explored in detail, including their invention, function, and technical peculiarities. You'll also find information about components found on buses, motorcycles, bicycles, and more. Even seasoned gearheads will learn from this guide as it traces the history and development of mechanisms they may take for granted.

978-1-55652-812-5
$14.95 (CAN $16.95)

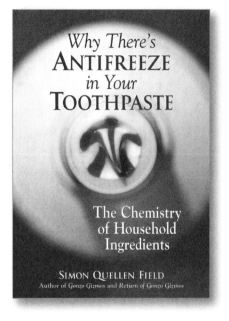

978-1-55652-697-8
$16.95 (CAN $18.95)

Why There's Antifreeze in Your Toothpaste

The Chemistry of Household Ingredients

Simon Quellen Field

If you're like most people, you find it hard enough to *pronounce* the ingredients found in most household products, much less understand why they're there. No longer—with *Why There's Antifreeze in Your Toothpaste* you'll be able to distinguish between preservatives and sweeteners, buffers and emulsifiers, stabilizers and surfactants. Ingredients are grouped according to type, and each entry contains the substance's structural formula, synonymous names, and a description of its common uses. This helpful guide can be used as a basic primer on commercial chemistry or as an indexed reference to specific compounds found on a product label. Never fear the grocery again.

978-1-55652-651-0
$12.95 (CAN $16.95)

Skewered!

The Rudest Food Reviews

Compiled by Michelle Lovric

Ever since Goldilocks made saucy remarks about the temperature of her porridge, people have been saying nasty things about other people's cooking. From "the food would create an insurrection in the poorhouse" (Mark Twain) to "the coffee tastes like water that has been squeezed out of a wet sleeve" (Fred Allen), *Skewered!* compiles the most vile morsels of wit ever written by malicious gourmets. Take this handy companion to your least favorite restaurant, and you'll never be stuck for dinner conversation again.

CHICAGO
REVIEW
PRESS

Available at your favorite bookstore, by calling
(800) 888-4741, or at www.chicagoreviewpress.com